The Unpredictable Species

Philip Lieberman

The *Un*predictable Species

WHAT MAKES HUMANS UNIQUE

PRINCETON UNIVERSITY PRESS

Princeton and Oxford

Library of Congress Cataloging-in-Publication Data
Lieberman, Philip.
  The unpredictable species : what makes humans unique / Philip Lieberman.
      pages   cm
  Includes bibliographical references and index.
  ISBN 978-0-691-14858-8 (cloth : alk. paper)   1. Brain—Evolution.   2. Human
evolution.   3. Evolutionary psychology.   I. Title.
  QP376.L563 2013
  612.8′2—dc23   2012037744

British Library Cataloging-in-Publication Data is available

To Marcia

Contents

Preface

Scarcely a day passes without a report about genes associated with debilitating diseases such as early onset diabetes or breast cancer. Other genes allow you to happily eat ice cream and drink milk because Darwinian natural selection acted on your distant ancestors, allowing them to digest lactose beyond infancy. And you can hardly be unaware of the debate surrounding genetically modified organisms (GMOs) in the food chain or on the farm. Many of the genetic differences between humans and chimpanzees are now clear. Advances in genetic research are producing startling results. Trading on these developments, it isn't surprising that a simple explanation for the way that we humans act has been accepted in academic circles and by most everyone else. We supposedly have genes that confer language, make us moral or religious, favor our close relatives, allow us to detect cheats, determine our taste in art, and so on. Simple solutions to complex problems are always appealing.

The focus here will be on research that points to understanding how the human brain evolved so as to allow us to choose between alternative courses of action and to create new possibilities—in short, the creative capacity that sets us aside from other species. Presently, no one can definitively tell you how the human brain works, but we can rule out the nineteenth-century models that guide the "gene that controls something" school of thought—whether it is language, math, morality, favoring relatives, and the like. Advances in neuroscience show that our brains do not operate in a manner similar to the components that make up digital computers. The biological brain doesn't consist of independent "modules"—subsystems that each control a specific function, one module operating the keyboard or touch screen, another devoted to the display, and so on. Instead, specific structures of the brain clearly carry out particular "local" operations, but these operations don't constitute an aspect of behavior that you can observe. The "instructions" for moving your arm forward, which involves a learning process carried out by a set of linked parts of the brain, are stored in the motor cortex, but even so seemingly simple a task as lifting a glass of water to your mouth involves pulling these instructions out of this motor memory store through local operations carried out in different parts of your brain, linked together in a "circuit." I will describe some of these neural circuits and what they do, peeling away later in the text, the technical jargon.

Surprisingly, brain structures whose roots can be traced back to before the age of dinosaurs play a critical role in circuits involved in learning and executing motor acts, as well as in thinking. Neural circuits that first evolved for motor control were tweaked so that they regulate a range of linguistic and cognitive acts, including our creative ability to form new ways of acting and thinking, and choosing among them. Charles Darwin wasn't aware of DNA, nor was he aware of the mechanisms of genetic evolution, but ongoing studies show that the guiding principles of evolution that Darwin proposed account for why we are not hairless chimpanzees, even though almost 99 percent of our genes are similar. Instead, natural selection acting on genetic variation, and the opportunistic use of existing organs adapted for one function to another, accounts for the neural bases of human language and our not being ruled by genes fixed in human prehistory or in our animal forbears.

The story isn't over—we're just at the starting point, but it is clear that determinist theories like those proposed by Noam Chomsky are

untenable. As we'll see, attested evidence of Darwinian natural selection acting on humans and the presence of genetic variation rule
out Chomsky's views, as well as related theories that claim that we
have "moral genes," "cheater-detector genes," or "monogamy genes."
Other living species show signs of creativity, and that certainly was
the case for extinct hominins such as the Neanderthals, but cultural
innovation fueled by the enhanced creative capacities of human
brains has made us the unpredictable species. The result is that we
are able to shape our actions and destiny for better or worse.

Philip Lieberman
Providence, Rhode Island
May 30, 2012

Acknowledgments

I must stress my debt to the work of hundreds of scientists over the course of more than a century. At a personal level, I owe much to my students and colleagues (the distinction has become very indistinct), who have continually pushed me to think harder about how we came to be and to carry out studies to test guesses (aka hypotheses). Terrance Nearey, and Robert Buhr were among the first. Molly Mack and Karen Landahl, whose untimely deaths have left a void, provided insights on how children and adults learn languages and the nature of speech and language, as did Judith Parker, Joan Sereno, Allard Jongman, and Patricia Kuhl. Joseph Friedman, Liane Feldman, Emily Pickett, Hiroko Nakano, Jesse Hochstadt, Beverly Young, Angie Morey, Sheila Blumstein, Jennifer Adeylott, Thanassi Protopapas, Sonja Kotz, John Donoghue, Ann Graybiel, and Oury Monchi helped me understand how brains seem to work. Alvin Liberman, Franklin Cooper, Edmund Crelin, Arend Bouhuys, Ken Stevens, Dennis Klatt, and Jeffrey Laitman played key roles in research on the role of speech in language and its evolution. Though I think that he has set a wrong course, Noam Chomsky initially set me on this journey. Tecumseh Fitch, before he signed onto Chomsky's crew, made significant contributions on the evolution of human speech. Chu-Yu Tseng guided me through the tone languages that most people on earth speak. Liz Bates and Fred Dick helped in everything. My son Daniel Lieberman, with Robert McCarthy, showed that the time course for the development of the anatomy that allows humans to produce sounds that enhance speech communication extends over the first eight years of life, instead of a year. Daniel's comprehensive study of the evolution of the human skull has been invaluable. My son Benjamin's book, *Terrible Fate: Ethnic Cleansing in the Making of Modern Europe*, forcefully pointed out facts that cast doubt on any genetic basis for moral conduct. Robert McCarthy's studies of the fossil record of human evolution were especially helpful. Svante Paabo graciously shared his thoughts on the transcriptional factor, FOXP2, that distinguishes us from other species. Eric Schwartz at Princeton University Press provided sagacious editorial advice throughout this project. And Marcia Lieberman put up with my continually intruding on her own work to see what she thought of what I was thinking of at the moment. But any misinterpretations of anyone's research or ideas are my own.

The Unpredictable Species

Chapter One

Brainworks

My job was to alert the driver to the wood-framed carts pulled by one or two bullocks, usually one. None of the carts had tail lights, or for that matter, any lights! It was raining—monsoon time. We had been driving for about ten hours to New Delhi from the foothills of the Indian Himalaya, where we had walked for a month from one monastery to another, photographing Tibetan wall paintings. The wipers were oscillating, but the windshield was greasy. On the short section of "high-speed highway," dusk was falling and the carts were becoming invisible. Sandwiched between the Tata trucks belching black exhaust fumes and intercity buses hurtling past with horns blaring were examples of Indian ingenuity—moving installations that could have been at home either in the Whitney Biennial for contemporary art or in a junkyard. The strangest was a huge tricycle. A massive steel I-beam, probably scavenged from a bridge, connected a giant front wheel to two small rear wheels. At the front, the driver straddled the beam about eight feet above the road. Behind him a metal box, attached to the beam, slanted downward, crammed with men, women, and children. The engine hanging below the steel looked as though it might have once been connected to a village water pump. The operating design principle was simple: use anything you can find to make it work.

The design principles that guided the evolution of the human brain are similar. Ernst Mayr, who shaped twentieth-century evolutionary and genetic research, repeatedly pointed out the "proximate logic of evolution." Existing structures and systems are modified sometimes elegantly, sometime weirdly, to carry out new tasks. Our capacity for innovation, which distinguishes human behavior from that of any other species, living or extinct, is a product of this minimal-cost design logic.

The path of human evolution diverged from chimpanzees, our closest living relatives, five to seven million years ago. The brains of our distant ancestors had started to enlarge a million years ago, but big brains alone don't account for why we act and think in a manner that differs so radically from chimpanzees. We are far from a definitive answer, but converging evidence from recent genetic, anatomical, and archaeological studies shows that neural structures, which have an evolutionary history dating back to when dinosaurs roamed the planet, were modified ever so slightly to create the cognitive flexibility that makes us human. Apart from our ability to acquire a vast store of knowledge, we have a brain that is supremely capable of adapting to change and inducing change. We continually craft patterns of behavior, concepts, and cultures that no one could have predicted.

The archaeological record and genetic evidence suggest that people who had the same cognitive capabilities as you or me probably lived as far back as as 250,000 years ago. However, we don't live the way our distant ancestors did 50,000 years ago. Nor do we live as our ancestors did in the eighteenth century, or five decades ago. Nor, for that matter, does everyone throughout the world today act in the same manner or share the same values. Unlike ants, frogs, sheep, dogs, monkeys or apes—pick any other species other than *Homo sapiens*—our actions and thoughts are unpredictable. We are the unpredictable species.

The opposite view, popularized by proponents of what has come be known as "evolutionary psychology," such as Noam Chomsky, Richard Dawkins, Sam Harris, Marc Hauser, and Steven Pinker, is that we are governed by genes that evolved in prehistoric times and never changed thereafter. The evidence presented to support their theories often involves colorful stories about life 50,000 or 100,000 years ago, sometimes supplemented by colorful blobs that are the end-products of expensive functional magnetic resonance imaging (MRI) machines can monitor brain activity. The red or yellow blobs are supposed to reveal "faculties"—specialized "centers" of the brain that confer language, altruism, religious convictions, morality, fear, pornography, art, and virtually everything else. I shall show that no one is religious because a gene is directing her or his beliefs and thoughts. Moral conduct doesn't entail having a morality gene. Language doesn't entail having a language gene.

The current popular model of the brain is a digital computer. It used to be a telephone exchange; at one time, it was a clock or steam engine. The most complex machine of the day usually serves to illustrate the complexity of the human brain. No one can tell you how the brain of even a fly, frog, or a mouse works, but the research that I will review shows that the software, computational architecture, and operations of a digital computer have no bearing on how real brains work. Your brain doesn't have a discrete "finger-moving" module that conceptually is similar to the electronics and switches that control your computer's keyboard. Nor do computer programs bear any resemblance to the motor control "instructions" coded in your brain that ultimately move your finger.

As Charles Darwin pointed out, in the course of evolution structures that originally had one purpose took on new roles and were modified to serve both old and new tasks. It has become evident that different neural structures, linked together in "circuits," constitute the brain bases for virtually every motor act and every thought process that we perform. We have adapted neural circuits that don't differ to any great extent from those found in monkeys and apes to perform feats such as talking, dancing, and changing the way that we act to one another, or to the world about us. The whole story isn't in. However, the studies that I will review have identified some of the genes and processes that tweaked these circuits, to yield the cognitive flexibility that is the key to innovation and human unpredictability.

The Functional Architecture of the Brain

Your car's shop manual provides a better guide to understanding how your brain works than books such as Steven Pinker's 1998 *How the Mind Works*. If your car doesn't start, the shop manual won't tell you to replace its "starting organ" or "starting module"—a part or set of parts dedicated solely to starting the car. The shop manual instead will tell you to check a set of linked parts: the battery, starter relay, starter motor, starter switch, perhaps the transmission lock. Working together, the "local" operations performed in each part start the car. Your car doesn't have a "center" of starting. Nor will you find a localized electric power "module." The battery might seem to be the ticket, but the alternator, powered by the engine, controlled by a voltage regulator, charges the battery. And if you're driving a gas-electric hybrid, the brakes also charge the battery when you slow down or stop. The brakes, in turn, depend on the electrical system because they're controlled by a computer that monitors road conditions, and the computer needs electric power. The engine is controlled by an electrically powered computer that senses the outside temperature and engine temperature to control fuel injectors and the timing of the electrical ignition of the fuel injected into each cylinder.

Each component—the battery, voltage regulator, fuel pump, ignition system—performs a "local" operation. The linked local operations form a "circuit" that regulates an observable aspect of your car's "behavior"—whether it starts and how it accelerates, brakes, and steers. And the local operation performed in a given part can also play a critical role in different circuits. The battery obviously powers the headlights and the sound system, as well as the fuel pump, ignition system, starting circuit, and the computer that controls the engine. The individual components often carry out multiple "local" operations. The engine can propel or brake. The car manual advises downshifting and using the engine to brake when descending long, steep roads. Some circuits include components that play a critical role in other circuits. In short, your car wasn't designed following current "modular" theories that supposedly account for how brains work and how we think.

Modularity

In contrast, according to practitioners of evolutionary psychology, various aspects of human behavior, such as language, mathematical capability, musical capability, and social skills, are each regulated by "domain-specific" modules. Domain-specificity boils down to the claim that each particular part of the brain does its own thing, independent of other neural structures that each carry out a different task or thought-process. Steven Pinker, who adopted the framework proposed in earlier books by Jerome Fodor (1983) and Noam Chomsky (1957, 1972, 1975, 1980a, 1980b, 1986, 1995), claims that language, especially syntax, derives from brain mechanisms that are independent from those involved in other aspects of cognition and, most certainly, motor control. Further refinements of modular theory partition complex behaviors such as language into a series of independent modules, a phonology module that produces and perceives speech, a syntax module that in English orders words or interprets word order, and a semantics module that takes into account the meaning of each word, allowing us to comprehend the meaning of a sentence. These modules can be subdivided into submodules. W.J.M. Levelt (1989), for example, subdivided the process by which we understand the meaning of a sentence into a set of modules. Each hypothetical module did its work and sent the product on to another module. Levelt started with a "phonetic" module that converted the acoustic signal that reaches our ears to segments that are roughly equivalent to the letters of the alphabet. The stream of alphabetic letters then was grouped into words by a second module that has no access to the sounds going into the initial phonetic module, the words then fed into a syntax module, and so on.

We don't have to read academic exercises to understand modular architecture. Henry Ford's first assembly line is a perfect example of modular organization. One workstation put the hood on, another the doors, another the wheels and tires, and so on. Each workstation was "domain-specific." The workers and equipment that put doors on the car frame put on doors, nothing else. Another crew and equipment put hoods onto the car frame. The assembly line as a whole was a module devoted to building a specific type of car. In Henry Ford's first factory, the Model T was available in any color, so long as it was black. As the Ford Motor Company prospered, independent "domain-specific" assembly lines were opened that each built a particular type of car or truck. But the roots of modular theory date back long before, to what neuroscientists generally think of as quack science, "phrenology."

Phrenology

Phrenology, when it's mentioned at all in psychology or neuroscience texts, usually is placed in the same category as believing in little green people on Mars. But in the early decades of the nineteenth century, phrenology was at the cutting edge of science. Phrenology attempted to explain why some people are more capable than others when it comes to mathematics. Why are some people pious? Why are some people greedy? Why do some people act morally? The answer was that some particular part of our brain is the "faculty" of mathematics, morality—whatever— and is responsible for what we can do or how we behave.

Whereas present-day pop science links these neural structures to genes that ostensibly result in specialized neural "faculties" located in different modules in discrete parts of your brain, phrenologists had a simpler answer—your abilities and disposition could be determined by measuring bumps on your skull. Phrenologists thought that they could demonstrate that a particular bump was the "seat" of the faculty of mathematics, which conferred mathematical ability; a different bump was the seat of the language faculty; another bump was the seat of the moral faculty; and so on. The size of each bump determined the power of the faculty. A larger bump would correlate with increased capability in language if it was the seat of the faculty of language. A larger bump for the seat of the faculty of piety would presumably be found on the skulls of clerics known for piety. A larger seat of mathematics would characterize learned mathematicians, and so on. The proposed faculties were all domain-specific—

independent of each other. The bump signaling piety, for example, had no relevance to language. This theory may sound familiar when transmuted into current modular theory, which pushed to its limits claims that you can be a mathematical genius but can't find your way home from work.

The labeled "map" in figure 1.1, from Johann Spurzheim's (1815) phrenological treatise, shows the seats of various abilities and

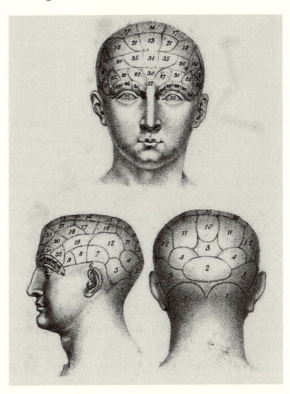

Figure 1.1. Phrenological maps of the skull showing the "faculties" of behavior and cognitive capacities. Early nineteenth-century phrenological maps showed the "seats" (areas and bumps on a person's skull) of faculties that supposedly determined various attributes such as morality, violence, piety, and so on, as well as language, mathematical ability, artistic ability, and other skills. The area of each seat indicated the degree to which an individual had a particular trait. One of the principles of phrenology was that the neural operations in each seat were specific to each faculty, an assumption that still marks many current mind/brain/language theories. From Spurzheim, J. G, *The Physionomical system of Drs. Gall and Spurzheim*, 1815.

personality traits. In the 1970s, I found skulls that had been carefully engraved with phrenological maps in the storage areas of the Musée de l'Homme in Paris. Phrenology collapsed because it was open to test. The skulls of clerics, leading mathematicians, musicians, homicidal maniacs, and so on, were measured. There was no correlation between the size of the bumps on a person's skull and what she or he could do or how she or he behaved. Some homicidal maniacs had bigger moral bumps than worthy clerics. And so phrenology was declared dead.

Pop Neuroscience

But phrenology lives on today in studies that purport to identify the brain's center of religious belief, pornography, and everything in between. The research paradigm is similar to that used by phrenologists 200 years ago, except for the high-tech veneer. Typically, the brains of a group of subjects are imaged using complex neuroimaging systems (described later) that monitor activity in a person's brain while he or she reads written material, answers questions, or looks at pictures. The response seen in some particular part of the brain is then taken to be *the* neural basis of the activity in question.

In one such study published in 2009, "The Neural Correlates of Religious and Nonreligious Belief," Sam Harris and his colleagues monitored the brain activity of 15 committed Christians and 15 nonbelievers who were asked to answer "true" or "false" to "religious" statements, such as "Jesus Christ really performed the miracles attributed to him in the Bible," and "nonreligious" statements such as "Alexander the Great was a very famous military leader." Activity in an area in the front of the brain, medial ventrolateral prefrontal cortex, was greater for everyone when the answer was "true." This finding is hardly surprising since, as we will see, ventromedial prefrontal cortex is active in many cognitive tasks, including pulling information out of memory (Hazy et al., 2006; Postle, 2006). Deciding that something is true also takes more effort when the task is pushing a button when you think that a word is a real English word, for example, "bad," than not pushing the button when you hear "vad" (Rissman et al., 2003). In fact, ventromedial prefrontal cortex is active in practically every task that entails thinking about anything (Duncan and Owen, 2000).

But the neural structures that Sam Harris and his colleagues claimed were responsible for religious beliefs included the anterior insula, a region of the brain that the authors themselves noted is also associated with pain perception and disgust; the basal ganglia, which I'll discuss in some detail because it is a key component of neural circuits involved in motor control, thinking, and emotion; and a brain structure that all mammals possess—the anterior cingulate cortex (ACC). The ACC supposedly was the key to religious belief because it was more active when the committed Christians pushed the "true" button about their religious beliefs. The ACC is arguably the oldest part of the brain that differentiates mammals from reptiles, and dates back 258 million years ago to the age of dinosaurs. As we'll see, its initial role probably was mother-infant care and communication (MacLean, 1986; MacLean and Newman, 1988). Sam Harris's agenda is scientific atheism—religious belief supposedly derives from the way that our brains are wired. All mammals have an ACC. If Harris and colleagues are right, mice may be religious!

Increased ACC activity often signifies increased attention. The ACC activity observed by Harris et al. (2009) in the committed Christians may have reflected the subjects' paying more attention to the emotionally loaded questions probing their faith or lack thereof. An obvious control condition would have monitored the subjects' responses to questions about their tax returns.

Another recent exercise in pop neuroscience "explains" why young men are more interested in pornography than women. In a study published in a first-line journal, *Nature Neuroscience*, Hamann et al. (2004) showed erotic photographs to 28 Emory University male and female undergraduates. The neuroimaging apparatus, which costs several million dollars, showed that the men had more activity in two closely connected brain structures, the amygdala and hypothalamus. The authors, unsurprisingly, conclude that male undergraduates may be more interested in erotic photographs

9

than women. A glance at a magazine stand near Emory would have reached the same conclusion, saving several thousand dollars spent on the experiment. The main finding of the study is that increased amygdala activity, in itself, accounts for "why men respond more to erotica." The conclusion is weird; dozens of independent studies show that activity in the amygdala increases in fearful situations. Amygdala activity increases, for example, when people look at angry faces. Were the men afraid of sex? However, other experiments show that the amygdala also is activated by stimuli associated with reward (Blair, 2008), so the pictures may have reflected wishful thinking. What would the amygdala responses have been if the men had been imaged while they looked at expensive sports cars or alternatively at fearful faces? What would have happened if the erotic pictures had been shown to the "committed Christians" studied by Sam Harris and his colleagues?

We'll return to pop-neuroscience "proofs" for morality and other qualities in chapter 6.

A Pocket Guide to the Human Brain

Most of the human brain is divided into two almost equal halves, or cerebral hemispheres. Figure 1.2 shows a side view of the outside of the left hemisphere of the human brain. The neocortex is the outermost layer of the primate brain. In this sketch of the left surface of the neocortex, orbofrontal prefrontal cortex is at the extreme front. Dorsolateral prefrontal cortex is sited behind it in the upper part of the cortex. Ventrolateral prefrontal cortex is below it. (The Latinate terminology simply refers to the location of each cortical area.) Figure 1.3, shown later, will also be useful in locating these "areas" of the cortex. The brains of mammals, other than aquatic dophins and whales, are marked by "fissures," deep folds separating "gyri." Some of the external landmarks of the brain that can be seen using current techniques that monitor brain activity are pictured. The Sylvian fissure, the deep channel that is the boundary between temporal lobe and frontal lobe of the cortex, for example, shows up clearly. The

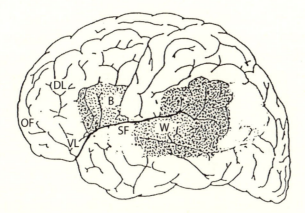

Figure 1.2. The left side of the human cortex. The "anterior," prefrontal regions of the cortex are to the left. Orbofrontal prefrontal cortex is located at the marker OF, dorsolateral prefrontal cortex at the marker DL, and ventrolateral prefrontal cortex at the marker VL. These prefrontal regions are involved in language and a range of cognitive tasks; they vary in size and position from one person to another. The Sylvian fissure, SF, marks the boundary between frontal and posterior cortex and is visible in neuroimaging studies. The traditional language "organs" of the brain also are mapped. Although Broca's and Wernicke's areas can be involved in language tasks, they are not the brain bases of language. They can be destroyed without permanent language loss; other areas of the brain perform critical roles, making both human language and cognition possible.

human neocortex and most other parts of our brain are about three times larger than a chimpanzee's.

Neuroimaging

Neuroimaging techniques that visualize the brains of living persons and track brain activity while people think have led to a revolution in human brain research. But the limitations of current neuroimaging techniques have to be understood to sort out the claims of pop neuroscience from real science.

Computed tomography (CT) scans, computer-derived three-dimensional x-ray images of the brain, first provided images of the brains of living patients. Magnetic resonance imaging (MRI) then came into wide use in the mid-1980s. MRI, which reveals the soft tissue structure of the brain, is the keystone of all neuroimaging techniques, because if you don't know where brain activity is occurring, you won't have a clue to what's happening, other than that the subject is alive.

The heart of the MRI apparatus is a powerful electromagnet—that's why there are cautions when entering the MRI room about metal fragments in your eyes or having objects on your person that contain iron or steel. The magnet can tear metal fragments loose and pull metal objects into the apparatus. A strong magnetic field first displaces the electrons of the atoms that form the soft tissue of the brain from their rest position while the subject is in the MRI apparatus. The magnetic field produced by the electromagnet then is abruptly turned off, causing the electrons to snap back to their rest position. This produces an electrical signal at frequencies that complex computer programs use to determine the composition of the atoms.

If the magnet is the MRI's heart, its brain is its computer. The computer programs essentially produce a series of images of thin virtual "slices" of a person's brain. The person in the MRI apparatus ("magnet" is MRI jargon) hears the noise produced by the magnetic elements in the magnet's iron core being switched on and off. The downside of any technique based on MRI is the noise—the series of sharp clicks that occur as the MRI apparatus produces a series of slices that are assembled to form a three-dimensional image of the brain. In the early days of MRI, the patient/subject was positioned lying in a narrow tube. An acquaintance who had served in the US Navy during World War II described the experience as being in a submarine's torpedo tube while hearing a machine gun firing. MRI and all techniques based on MRI inherently cannot track rapid events.

While MRI reveals the structure of the brain, including damage, functional magnetic resonance imaging (fMRI) and positron emission tomography (PET) track brain activity. If you wanted to track the amount of energy consumed by your car at any instant of time, you could monitor either the flow of fuel or the oxygen used to burn the fuel. The brain is a sophisticated engine. The fuel is sugar—

glucose. PET tracks glucose fuel consumption in different parts of the brain to infer the pattern of local neural activity. PET works by monitoring the distribution of radioactive glucose. It's necessary to first place a vial of glucose in a cyclotron to make a radioactive form of glucose, an "isotope" that will rapidly lose its radioactivity. The radioactive isotope has a short "half-life"—the time interval in which the radioactive level will diminish by half. The radioactive glucose is then injected into a volunteer's brain. The PET apparatus then tracks the injected glucose in different parts of the brain by monitoring the emission of subatomic particles. PET is useful because, unlike fMRI, no magnet noise occurs, but it has its drawbacks. It does not yield very detailed images, and there is a short working period between placing the glucose sample in a cyclotron (which isn't a standard laboratory instrument) and studying the subject's responses in an experiment.

fMRI cleverly solved some of these problems by monitoring the by-product of combustion—oxygen depletion. As parts of the brain become more active, they burn more glucose, using up more oxygen. fMRI tracks the relative level of local brain activity by tracking the local level of oxygen left after burning brain fuel. As glucose is burned, the oxygen level falls, reflecting brain activity. The computer system is adjusted to look at the electronic signature of the local oxygen level. The depleted blood oxygen level—the "BOLD" response—thus is the fMRI signature of local brain activity. Direct comparison with neural activity monitored with micro-electrodes inserted in the brains of animals confirms that the BOLD signal is a valid measure of neural activity (Logothetis et al., 2001). Direct recording of neural activity using inserted electrodes, with the exception of some highly limited circumstances, cannot be performed in human subjects.

One of the major problems in interpreting the signals coming out of a PET scanner or fMRI system's computer has to do with the primary reason for using these techniques—the person who's being studied is alive. Therefore, there is activity constantly going on in many parts of the subject's brain. One technique that has been used to attempt to find out whether the neural activity that has been recorded has anything to do with the behavior that the experiment is supposed to be studying is to subtract the signals recorded from a "baseline" from the "experimental" condition. In an experiment that's aimed at discerning the brain mechanism involved in listening to speech perception, the subjects might be asked to listen to "pure" musical-like tones. The presumed neural mechanisms involved in speech perception would then be determined by subtracting the activity recorded when the subjects listened to musical tones from the activity recorded when they listened to words. The problem that arises is that some neural mechanisms may be involved in listening both to musical tones and to speech, so when the subtraction is performed you don't see that they also are active when listening to words. That problem has plagued PET and fMRI studies since they first were used. In fact, recent work shows that many of the neural mechanisms used in speech are also used to perceive music (Patel, 2009). Another problem arises from using "region of interest" (ROI) computer processing that examines only neural activity in a predetermined region of the brain. If you don't look for brain activity in some part of the brain, you won't find it.

One way of addressing some of these problems is to use a task that can be performed at increasing levels of difficulty. If activities in particular neural structures increase with task difficulty, they most likely are relevant to the task. Information on pathways—the neural circuits that link different parts of the brain—also should be taken into account when brain imaging data are evaluated. We'll discuss the findings of studies that use these techniques in the chapters that follow.

Diffusion tensor imaging (DTI), another variation on MRI imaging, can trace the neural circuits that connect the structures of the brain in living humans (Lehericy et al., 2004). The principle behind DTI is amazingly simple. Suppose that you dip a stalk of celery into a bowl of water and then invert it. Water will flow down along the stalk. DTI tracks the water flow pattern throughout the brain's circuits because the flow of neurotransmitters follows the bundle of "white" matter linking the "gray matter" of the neural structures that make up the circuits.

Event-related potentials (ERPs) provide an indirect measure of rapid changes in neural activity within the brain. Electric currents, including the signals that transmit information in the brain, always generate magnetic fields. The magnetic field can be monitored using an array of small antenna-like coils placed on a person's skull. Although localizing the source of the radiated magnetic field to very specific parts of the brain is not possible, rapid changes can be discerned that correspond to cognitive acts. Angela Frederici, Director of Neuropsychology at the Max Planck Institute in Leipzig, Germany, in her 2002 review article, shows that the subcortical basal ganglia play a part in rapidly integrating the meanings of individual words with syntactic information as a subject listens to a spoken sentence.

Where Is What

There is an inherent problem that affects all neuroimaging studies. The "same" locations noted in different studies may not be the same. Devlin and Poldark, in their 2004 paper "In Praise of Tedious Anatomy," show that it's not necessarily the case that the hot-spots indicating increased activity in fMRI scans in a given study occur in the same parts of the brains of the many subjects who must be imaged to obtain usable data. The published fMRI brain maps using red, blue, and green to indicate relative levels of activity in different parts of the brain are derived by complex computer-implemented processing. fMRI signals are weak, and it's necessary to monitor many subjects performing the same task and then average the responses to factor out irrelevant electronic noise. But "averaging" these responses is not simple. People's brains differ from one to another as much as their noses, hands, hair, arms, legs, faces, and so on. Because the shape and size of the brain varies from person to person, the MRI brain map of each subject must be stretched and squeezed in different directions until it conforms to a "standard" brain, often the Talairach and Tournoux (1988) coordinates, that attempt to reshape all brains to that of a single 60-year-old woman. The problem is especially acute for frontal and prefrontal cortical areas because they don't have sharp boundaries that show up in the structural MRI that the fMRI activity is mapped onto.

The net result is that it is often impossible to be certain that the hot-spots actually are in the same area of the cortex. Devlin and Poldark proposed using different brain mapping techniques that may provide more realistic brain maps, but some uncertainty always remains, impeding precise localization of brain activity. The level of uncertainty paradoxically increases in studies that attempt to be extremely precise. The brain maps proposed by Korbinian Brodmann between the years 1908 and 1912 have been used to identify the sites of brain damage or neural activity in thousands of studies. The basic "atomic" computational elements of the brain and nervous system are "neurons." Neurons come in different sizes and shapes, and the neocortex has six layers of neurons arranged in columns. Brodmann microscopically examined human brains and found differences in the "cytoarchitecture"—the distribution of neurons in the columns that made up different parts of the neocortex, which he thought reflected functional differences. He also examined the brains of other animals and was also able to show that homologies—similarities in cytoarchitecture—existed between the brains of closely related species.

Figure 1.3 shows Brodmann's 1909 maps of the left cortical hemispheres of a rhesus monkey and a human. Some of the Brodmann areas (BAs) are easy to pick out on an MRI. Area 17, at the extreme tail end of the brain, is one. In contrast, precisely locating the BAs that constitute frontal and prefrontal cortex on an MRI is flaky. One area merges seamlessly into another in the MRI. As Devlin and Poldark point out, if you really want to locate a frontal cortical Brodmann area, it's necessary to carry out a microscopic examination of the subject's cortex. This would entail sacrificing the subjects. In an exercise aimed at seeing whether it is possible to precisely identify Broca's area, which consists of BAs 45 and 44, the traditional "language organ" of the human brain (chapter 2 will show that Broca's area is not the brain's language organ), Amunts et al. (1999) examined the brains of 10 persons from autopsies performed within 8 to 24 hours after death. Thousands of profiles of cell structure were obtained using an automated observer-independent procedure to construct three-dimensional cytoarchitectural reconstructions of

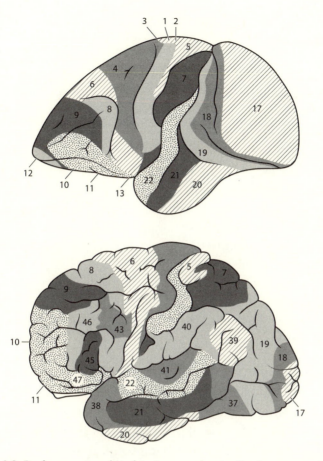

Figure 1.3. Brodmann cytoarchitechtonic maps, showing Brodmann areas (BAs) of the left cortical surfaces of a rhesus macaque and a human. Cortical areas having similar cytoarchitectonic structures mark all primates. However, the size and location of a given BA varies from individual to individual and generally cannot be identified with certainty in a living subject. BA 45 and BA 44 (the BA area to the right of BA 45) are the traditional sites of Broca's area. Ventrolateral prefrontal cortex includes BAs 47and 12. Dorsolateral prefrontal cortex includes BAs 49 and 9.

the individual brains. Both cytoarchitecture and the features of the brain that can be seen on an MRI varied from one individual brain to another to the degree that it is impossible to discern the location of either BA 44 or BA 45 from an MRI. Amunts and her colleagues also found 10-to-1 differences in the size of area BA 44 in their small

17

10-brain sample, close to the minimum number of subjects typically imaged in an fMRI study.

Cortical Malleability

Apart from uncertainty in identifying BA from MRIs, the differing cytoarchitecture of Brodmann areas is not an infallible guide to their function. For example, the parts of the cortex that usually are involved in vision take on different roles in blind people. The visual cortex in persons who were blind early in life is active when they read, "tactilely discerning shapes by touching Braille text and hearing speech sounds (Burton, 2003). Experience shapes cortical function, accounting for blind people using "visual" cortex to process tactile and auditory information. Research using animals strikingly shows the plasticity of the cortex. A series of experiments over 10 years involved rewiring the brains of ferrets. A research team at MIT directed by Mriganka Sur reversed the pathways leading from the ferrets' eyes and ears to their usual targets in visual and auditory cortex, shortly after birth (Melchner et al., 2000). When the ferrets reached maturity, direct electrophysiologic recording, which involved recording electrical activity from electrodes inserted into the animals' brains, demonstrated that they saw objects using cortical areas that normally would be part of the auditory system. Cortical malleability may explain why stroke patients who initially have language problems recover if the brain damage is limited to the cortex, noted by Stuss and Benson in their 1986 book directed at neurologists.

Neurons and Learning about the World

Neurons are the basic elements that make up the nervous systems of all animals. Figure 1.4 shows the cell body of a neuron, its output axon, and a profoundly reduced set of "dendrites" and synapses. I won't go into the details of how neurons signal to each other, but the discussion here should suffice to understand the significance

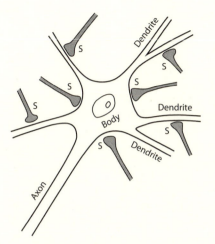

Figure 1.4. Neurons and synapses. The dendrites (the term in Latin refers to tree branches) branching out from each cell body connect neurons to each other. A cortical neuron typically receives inputs on these dendrites from several thousand other neurons. Each neuron has an output axon, which typically transmits information to thousands of other neurons. Incoming axons from other neurons transmit information to synapses on the dendrites as well as on the cell body.

of new insights from current, ongoing genetic research that will be discussed in chapter 4. It is becoming evident that at least one newly discovered gene makes our neurons more interconnected and more malleable, allowing us to free our behavior from the constraints that hobble apes.

Synapses are malleable structures that transfer information to a greater or lesser degree to neurons. They also are memory "devices." As a loose analogy, think of a synapse as the neural equivalent of the volume control of an audio amplifier. The degree to which the signal from the tuner, CD, MP3 player, or whatever gadget you've connected to the amplifier will be sent on to the amplifier depends on the setting of the volume control. If you set the volume control to a particular level, turn the system off, and turn it on the next day, the volume control's setting will constitute a memory "trace" of the sound level that you set on the previous day. Each synapse's "weight" (its volume control setting) thus constitutes a memory trace of the extent to which the incoming signals will be transmitted to the neuron. The neuron's output axon will "fire," producing an "action potential," an electrical signal that is transmitted to other neurons, when the signals transmitted from one or more dendrites and axons, added together, reach a threshold. Some synapses result in incoming signals having less effect in triggering an outgoing action potential.

The process by which synapses are modified is generally thought to be the basis for associative learning—the way that we and other animals learn about the world, and how we learn about our place in the world. Hebb in 1949 proposed that synapses are modified by the signals they transmit. Think of a water channel cut out of sandy soil. When water from two sources simultaneously flows through the channel, it will become eroded to a greater extent than if water were flowing through only one channel. Hence, the water channel subsequently will offer less resistance to the flow of water. The degree to which the channel becomes wider in a sense marks the degree to which the two sources are associated and thus constitutes a "memory trace" of the correlated events. If the channel is deeper and wider, more water can flow through it. Hebb suggested that the activity of neurons becomes correlated as they continually respond to stimuli that share common properties. Experiments that make use of arrays of microlectrodes that can record electrical activity in or near neurons confirm Hebb's theory. Enhanced conduction across synapses occurs when animals are exposed to paired stimuli (for example, Bear, Cooper, and Ebner, 1987), similar to the bells and meat used in Pavlov's classic experiments.

A Mollusk Academy

The classic example of associative learning taught in Psychology 101 involves dogs. Pavlov taught dogs to associate a bell's ringing with a meat treat. The bell was rung before the dogs were fed. Repeated "paired" exposure to the sound of a bell followed by meat resulted in a "conditioned reflex"—the dogs salivated. It's curious to note that some people also might salivate or automatically say "yum yum" or "that looks great" as they read a menu, but we wouldn't characterize their responses as "conditioned reflexes." When animals learn something, the overt behavior that we use to monitor their response is a conditioned reflex. When humans respond, they're thinking.

Pavlov thought that associative learning occurred in the cortex, but associative learning deriving from synaptic modification isn't limited to the cortex. A series of experiments in the early 1980s (Carew, Walters, and Kandel, 1981) showed that synaptic modification is the key to associative learning in mollusks. The curriculum of their mollusk academy wouldn't be approved by any school board. The mollusk species, *Aplysia californica*, apparently normally likes the taste/odor of extract of shrimp, stretching the meaning of the word *like* to animals that don't have a central nervous system. Twenty mollusks were trained using Pavlov's procedure. Essence of shrimp was presented to the mollusks for 90 seconds. Six seconds after the start of the presentation, an electrical shock was directed to the mollusks' heads. Six to nine paired shrimp-essence electric shocks were administered. Twenty other mollusks who served as a control group received random electric shocks. Eighteen hours later, shrimp extract was applied to all the mollusks as well as a weak electric shock to their tails. The "trained" mollusks ran away more rapidly, emitting ink (their escape ink-screen) and not feeding as well as the untrained (unpaired) mollusks. Critically, changes in the synapses in the motor-control neurons in their tails were observed. Synaptic modification was the neural mechanism by which the mollusks "learned" that essence of shrimp "means" that "I will be shocked."

Neural Circuits — Linking Cortex and the Rest of the Brain

However, it is not a mollusk that is tapping at the keyboard, spelling out the words that you are reading. It has become apparent that neural circuits, akin to the linkages between the parts of your car, link local operations throughout our brains, enabling us to walk, run, talk, think, and carry out the acts and dispositions that define our species. We can be acrobats or clods, saints or monsters — these possibilities derive from the neural circuits of our brains.

Diffusion tensor studies are identifying hundreds of human neural circuits. We will discuss the role of some of these circuits in motor control, vision, memory, and other aspects of cognition. Our focus will be on a class of neural circuits involving cortex and the subcortical basal ganglia, which are a set of closely connected neural structures deep within your brain. Contrary to the modular approach of modern-day phrenology, the cortical and subcortical elements that make up these circuits are not domain-specific—they make it possible for you to walk, focus on a task, talk, form or understand a sentence, or decide what you'll have for dinner or do with the rest of your life. In short, they are involved in motor control, language, thinking, and emotion. Hundreds of independent studies over the course of more than thirty years have established the local operations performed in these circuits, some of the aspects of behavior that they regulate, and why they differ from similar circuits in our close primate relatives. It has become apparent that they have a singular role in enhancing associative learning and cognitive flexibility. As you will see, the absence of modularity, neural structures committed to one aspect of human behavior, reflects their evolutionary history.

The basal ganglia evolved eons ago in anurans—the class of animals that includes present-day frogs (Marin, Smeets, and Gonzalez, 1998). They control motor acts in frogs and continue to do so in human beings, but they also play a central role in human cognition. Earlier theories on how the human brain works assumed that subcortical structures, such as the basal ganglia, were involved solely with motor control or emotion. You can still read or hear reports on CNN or NPR that claim that the brain's "pleasure center" is the basal ganglia. The cortex was supposed to control rational conduct, overriding subcortical emotional responses. Paul MacLean's 1973 Triune brain theory, for example, suggested that disorders such as Tourette syndrome, which causes a person to uncontrollably grunt or utter profanities, reflected the cortex losing control over the "emotional" subcortical brain.

That's not the case. Neural circuits that connect cortical areas with the basal ganglia and other subcortical structures do regulate emotion. However, other cortical-basal ganglia circuits make it possible to pay attention to a task. The ACC activity that was Sam Harris's brain marker for religious belief is part of this circuit. Still other circuits link prefrontal medial ventrolateral cortex and other prefrontal cortical structures with the basal ganglia and posterior regions of the cortex. These circuits are involved in comprehending the meaning of a sentence, and the range of cognitive acts that fall under the term "executive control" (for example, Badre and Wagner, 2006; Heyder et al., 2004; Lehericy et al., 2004; Miller and Wallis, 2009; Monchi et al., 2001, 2006b, 2007; Postle, 2006; Simard et al., 2011). Executive control is, to me, an odd choice of words for designating cognitive acts such as changing the direction of a thought process, or keeping a stream of words in short-term memory to derive the meaning of a sentence, subtracting numbers, and performing other cognitive tasks. What may come to mind is a domineering boss telling everyone what they must do or be fired, but we're stuck with the term.

Coming Attractions

It will become clear in the chapters that follow that apart from the size of the human brain, neural circuits involving cortex, basal ganglia, hippocampus, and other neural structures form the engine that drives human cognitive capabilities, including language. Genetic events that occurred in the last 200,000 to 500,000 years increased the efficiency of basal ganglia circuits that confer cognitive flexibility—the key to innovation and creativity. These mutations also enhanced human motor control capabilities, allowing us to talk.

Only humans can talk. Dogs can learn to comprehend the simple meanings of words; apes can learn to use manually signaled words to communicate and can understand simple sentences. Parrots can imitate some words and form simple requests, but no other animal can talk. Speech plays a central role in language, allowing us to transmit information to one another at a rapid rate, freeing our hands and gaze. We don't have to look at each other while we communicate. In the course of human evolution, the demands of speech shaped the human tongue, neck, and head, and have compromised a basic aspect of survival. We can more easily choke to death when we eat, because our tongues have been shaped to enhance the clarity of human speech.

This permits us to estimate when humans had brains that could take advantage of tongues optimized for talking at the cost of eating. These same brain mechanisms allow us to sing, dance, and carry out the odd acts that distinguish human behavior from that of other species, including chimpanzees, who are almost genetically identical to us.

But the genes dreamed up over coffee and cookies—or other substances—by evolutionary psychologists don't exist. The historical record that shows how people really act, as well as the evidence that reaches our ears and eyes every day, has been ignored. When we, for example, compare the behavior of present-day Norwegians, Danes, and Icelanders with their Viking ancestors, it is evident that no moral gene exists or ever existed. Ethical, moral behavior is a product of cultural evolution. This stew of invented genes diverts our attention from real progress in understanding the interplay of culture and biology in shaping human behavior.

In brief, we will explore the neural and cultural bases of the human capacities that have transformed the world. We will examine the central claim of evolutionary psychology, that we are locked into predictable patterns of behavior that were fixed by genes shaped by the conditions of life hundreds of thousands of years ago. The human brain evolved in a way that enhances both cognitive flexibility and imitation, the qualities that shaped our capacity for innovation, other aspects of cognition, art, speech, language, and free will.

Chapter Two

Brain Design by Rube Goldberg

In my youth, the cartoonist Rube Goldberg "designed" machines in the spirit of the giant motorized tricycle glimpsed on the road to New Delhi—odd assemblages that achieved an action by a set of ill-sorted steps that no one possessed of design logic would elect to build (figure 2.1).

The functional organization of the human brain doesn't include a parrot, but it is just as improbable. Neural structures that date back to the age of dinosaurs have been adapted in a manner as weird as any of Rube Goldberg's machines. As I pointed out in chapter 1, our brains don't have the simple, neat organization early nineteenth-century phrenologists proposed, or the modular structure that evolutionary psychologists push. The functional architecture of the human brain instead reflects the opportunistic logic of evolution.

Neophrenology

You may be puzzled because earlier I stated that phrenology, which claimed that our brains contained language organs, morality organs, and math organs, was declared dead about 1850. But most expla-

Self-Operating Napkin by Rube Goldberg

Figure 2.1. As you raise spoon of soup (A) to your mouth, it pulls string (B), thereby jerking ladle (C), which throws cracker (D) past parrot (E). Parrot jumps after cracker, and perch (F) tilts, upsetting seeds (G) into pail (H). Extra weight in pail pulls cord (I), which opens and lights automatic cigar lighter (J), setting off skyrocket (K), which causes sickle (L) to cut string (M) and allow pendulum with attached napkin to swing back and forth, thereby wiping off your chin. Artwork Copyright © Rube Goldberg Inc. All Rights Reserved. RUBE GOLDBERG ® is a registered trademark of Rube Goldberg Inc. All materials used with permission.

nations for why we have the ability to talk, understand words, and compose and understand sentences all come from what amounts to a phrenological model; we supposedly have two parts of the cortex, Broca's and Wernicke's areas, that constitute the brain's "language organ."

My only previous reference to Broca's area was that it is *not* the brain's language organ. I didn't even mention Wernicke's area. So why do you keep reading or being told that Broca's and Wernicke's areas are the brain bases of human language? Broca's area is often the touchstone for determining which of our distant fossilized hominin ancestors (the term "hominin" signifies a primate in or close to the line of human descent) talked. Televised documentaries on human evolution may feature someone in a white coat holding up a fossil skull and pointing to where they believe they can detect traces of Broca's area, supposedly showing that the fossil could have talked. So how can they all be wrong?

As the previous chapter pointed out, the basic premise of phrenology is that a discrete part of the brain, in itself, is the neural basis for an observable aspect of human behavior such as mathematical ability, morality, or language. Early nineteenth-century technology precluded examining a living person's brain to determine the boundaries of the cortical area that was the "seat" of language, or art, or piety, so phrenologists instead measured the area of a discrete part of the skull, which reflected the size of the area of the brain beneath it. The seats were domain-specific. The part of the skull conferring mathematical ability had nothing to do with whether someone was linguistically gifted or not. Phrenology collapsed precisely because it was a testable — hence, scientific — theory.

Phrenology Resuscitated — The Broca-Wernicke Language Theory

But phrenology was resuscitated in 1861when Paul Broca published one of the first studies of a neural "experiment-in-nature." If you take as a given the phrenological model (that presupposes that brains are machines) in which a given part is the "seat" of some observable aspect of behavior such as talking, then taking out that part should preclude talking. This clearly was a "forbidden" experiment. Broca instead initially studied two patients who had extreme difficulties talking after brain damage inflicted by strokes. Similar experiments-in-nature became the primary source of information on how brains might work until the development 100 years later of neuroimaging techniques, like fMRI and PET and EEG (electroencephalography, recording electrical activity on the scalp that reflects neural activity).

Broca's study of patient "Tan" is the better-known experiment-in-nature. Tan, whose real name was Leborgne, was a 51-year-old man who had a series of neurological problems. Leborgne could control the intonation, the "melody" of speech. (A primer on how speech is produced follows shortly.) However, Leborgne could not produce any recognizable words other than the syllable "tan," and thus was described as patient "Tan." According to phrenology, the "seat" of language, the part of the brain that controlled language, was between the eyes. Broca didn't question the underlying premise of phrenology — that a specific part of the brain was the seat of language, but he thought that the seat of language was located elsewhere.

The patient died soon after Broca saw him. An autopsy showed damage to the surface of the left frontal lobe of the patient's brain. However, Broca limited his observations to the surface of Leborgne's brain instead of sectioning it and systematically determining the nature and extent of damage. A few months later, Broca examined a second patient who could speak only five words after suffering a stroke. An autopsy showed brain damage at approximately the same part of the surface of the brain as for Leborgne. Broca, following in the footsteps of phrenology, localized language to the part of the brain that has since been termed Broca's area.

However, the actual pattern of damage to both patients' brains does not support Broca's phrenological theory. The brains of both patients had been carefully preserved in alcohol, and 140 years afterward, high-resolution MRIs were performed (Dronkers et al., 2007). The MRIs showed that damage was not limited to the part of the brain identified by Broca as the seat of language. In both patients, massive damage had occurred to the neural circuits that link cortex with other parts of the human brain. The pattern of damage involved the structures that make up the basal ganglia, other subcortical structures, and the pathways that connect cortical and subcortical neural structures. The MRIs also showed that the brain structure commonly labeled Broca's area—the left inferior gyrus, Brodmann's areas 44 and 45 (see figures 1.2 and 1.3)—wasn't the cortical area that was damaged in Paul Broca's first two patients. Parts of the brain to the front (anterior) to it instead were damaged.

In 1874, Karl Wernicke studied a stroke patient who had difficulty comprehending speech and had damage to the posterior temporal region of the cortex. Wernicke, in the spirit of phrenology, decided that this area was the brain's speech comprehension organ. Since spoken language entails both comprehending and producing speech, Lichtheim in 1885 proposed a cortical pathway linking Broca's and Wernicke's area. Thus Broca's and Wernicke's cortical areas became the neural bases of human language, supposedly devoted to language and language alone. Doubts were expressed soon afterward about the localization of language to a single part of the brain. Marie (1926), for example, basing his conclusions on autopsies of stroke victims, proposed that Broca's syndrome involved damage to the basal ganglia. However, new life was given to the Broca-Wernicke theory by David Geschwind in a 1970 paper published in the prestigious journal *Science*.

Why the Broca-Wernicke Theory Is Wrong

In the hundred years before Geschwind published his paper, postmortem autopsies were necessary to establish the pattern of brain damage that was responsible for "aphasia," permanently losing some aspect of language. But just about the time that Geschwind's paper appeared in print, CT scans—three-dimensional x-ray images of a living person's brain—were coming into general clinical practice. Whereas autopsies of the brain are not common, thousands of CT scans of patients' brains became available. It soon became clear that aphasia did not result from damage limited to these cortical "language areas." The traditional Broca-Wernicke theory is wrong. The syndromes, the patterns of possible deficits, are real, but they do not derive from brain damage localized to these cortical areas. CT scans showed that patients recovered or had only minor problems when Broca's and Wernicke's areas were completely destroyed, so long as the subcortical structures of the brain were intact. Geschwind, who participated in these studies, revised his views on the viability of the Broca-Wernicke theory shortly before his untimely death.

Moreover, CT scans subsequently showed that the language deficits of Broca's syndrome didn't necessarily involve damage to Broca's area. Damage to subcortical structures that supposedly had nothing to do with language, leaving Broca's and Wernicke's cortical areas intact, caused the classic signs and symptoms of aphasia—speech production deficits as well as difficulty comprehending the meaning of a sentence. Margaret Naeser and her colleagues in a 1982 research paper described the language problems of patients who had suffered strokes that spared cortex altogether, but selectively damaged the basal ganglia and pathways to it. Aphasia, permanent loss of language, occurs when the neural circuits linking cortical areas through the basal ganglia, thalamus, and other subcortical structures are damaged. The current position expressed by Stuss and Benson in their 1986 book directed at neurologists, *The Frontal Lobes*, is that aphasia never occurs absent subcortical damage.

The problem with the Broca-Wernicke theory and all variants of phrenology is that they fail to recognize the fact that brains are the product of the opportunistic illogic of evolution. As Dronkers and her colleagues note, the neural operations underlying speech and language "involve large networks of brain regions and connecting fibres." Locating the site of brain damage is easy compared to determining whether it's really the sole cause of a problem. For example, if you cut the wires that supply a high-voltage impulse to the spark plugs, your car won't start. But no mechanic, probably no one having even a vague notion of how cars work, would think that the wires constituted the "seat" of starting. The wires are just one element of the complex circuits linking many parts. Each part performs a "local" operation that is necessary to have a functioning car. And as the troubleshooting section of your car's service manual will point out, a defective part can cause many different, seemingly unrelated, problems. Why then does the Broca-Wernicke theory survive? Its survival probably hinges on its being a simple theory. TVs differ from radios in that they have extra parts. No radio has a visual display Humans can do things that no chimpanzee can achieve. We, therefore, must have some different part in our brain. In that frame of reference, Broca's area accounts for our being able to talk. And since the extra part is biologically specified, there must be a human language gene (perhaps a set of genes) that shaped Broca's area. Similar arguments would point to our possessing moral genes, cheater-detector genes, and selfish genes.

Neural Circuits

My wife and I spent part of the summers of 1993 and 1994 in a remote region of Nepal close to its Tibetan border. We slept in a tent on the roof of the home of one of the rich men of the principal village of Mustang, Lo Manthang. Mustang is a culturally Tibetan part of Nepal that had been annexed by the Gorkha monarchy that created Nepal in the eighteenth century. Lo Manthang has two large fifteenth-century Tibetan Buddhist "gompas"—monastic temples whose inner walls were covered with intricate wall paintings. The paintings are among the few original surviving examples of that period. The paintings of other temples were defaced in Chinese-occupied Tibet, and we had been awarded a grant from the Getty Foundation to document the paintings. My photographic equipment, which included staging to reach some of the paintings that were 20 feet above the temple floor, had been loaded onto the backs of a string of ponies because there was no road to Lo Manthang. The ponies, descendants of the ponies that carried the Mongol warriors of Genghis Khan, were docilely led by our horseman, Nima Wangdi.

The photographs and text that resulted from our work in Mustang are on the Brown University website (http://dl.lib.brown.edu/BuddhistTempleArt) and on a DVD commissioned by the Getty Foundation. What we didn't anticipate was that five years later, we would be waiting in a pediatric neurologist's office at Rhode Island Hospital in Providence to hear his assessment of whether Nima Wangdi's daughter would ever talk. Through a series of the improbable events that mark one's life, we had brought his daughter, who had epileptic seizures, to Providence for medical treatment that was not available in Nepal. Lhakpa Dolma, a bright-eyed, alert five-year-old girl, also was having difficulty talking. The neurologist had examined the MRI of Lhakpa Dolma's brain. He assured us that she would be able to talk because her "language organs," the left cortical areas that were the sites of Broca's and Wernicke's areas, were intact.

Would that that had been the case! At age eighteen, Lhakpa Dolma still cannot talk; her cognitive ability has also deteriorated. So the short answer to whether Broca's and Wernicke's cortical areas constitute our brain's language organ unfortunately is no. But the full answer, which the previous chapter touched on, is that neural circuits linking activity in many different parts of the brain control speech production, comprehend speech, understand or compose a meaningful sentence, and carry out the other acts and thoughts that in their totality constitute our "language faculty." What was apparent in Dolma's MRI was damage to the pathways connecting cortex and the basal ganglia.

Cortical-Basal Ganglia-Cortical Circuits

Diffusion tensor imaging (DTI), which can map out neural circuits in a noninvasive manner, has revealed hundreds of circuits in the human brain, including circuits that connect different parts of the cortex. The functional roles of some circuits have been teased out, but our knowledge is imperfect. That's also the case for many of the local operations performed in the cortex. Decades of research have shown that various bits and pieces of visual processing are performed in different cortical areas. Some parts of the cortex seem to be sensitive to lines, others to angles, others to color, but what we see when we look at a cat is a cat, not a jumble of lines, angles, and colors. No one can tell you how the "local" operations, the bits and pieces of visual processing, are put together to form the image that you see. How these local operations are integrated to form images remains mysterious. It's as though we had a giant road map that didn't show where any road went to.

But over the past thirty years, the destinations of a few circuits have become evident. The discovery process began with experiments-in-nature that focused on aphasia—Paul Broca's original experiment-in-nature. Over the course of the twentieth century, it gradually became apparent that patients who had suffered brain damage that resulted in aphasia, permanent loss of some aspect of linguistic ability, had cognitive deficits. Kurt Goldstein, in his 1948 book, based on observations of patients over the course of decades, also characterized aphasia as a loss of the "abstract capacity." Aphasic patients lost cognitive flexibility, pointing to neural circuits that regulated both language and cognition. Other experiments-in-nature, studying the effects of Parkinson disease and instances of brain damage localized to the basal ganglia, revealed a class of circuits that link cortical areas and the basal ganglia. These cortical-basal ganglia-cortical neural circuits play a central role in regulating motor control (including speech), comprehending the meaning of words and sentences, thinking, as well as emotional control.

Imagine a Martian scientist faced with the problem of determining how your car worked. He could systematically destroy individual parts and keep track of what happened. But most people would object to this practice, so the Martian could fall back to studying experiments-in-nature, cars that had broken down. Other experiments-in-nature, studying the effects of Parkinson disease and other instances of brain damage, filled in more blank spots. Neuroimaging studies of both neurologically intact "normal" subjects and patients are filling in some of the questions about precisely what particular neural structures do, as well as posing new questions.

Neural circuits linking cortical areas through the basal ganglia provide the basis for many of the qualities that differentiate us from animals. The basal ganglia are buried deep within the human brain (figure 2.2). The Latin terminology, which attempted to describe the shape of

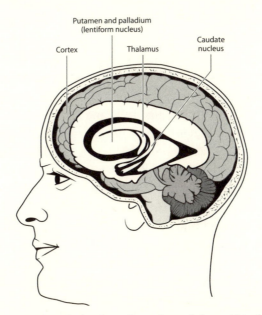

Figure 2.2. Basal ganglia. The basal ganglia are located deep within the skull. The putamen and palladium (another name for the globus pallidus) are contiguous and form the lentiform nucleus. The caudate nucleus and putamen form striatum. Cortical-to-basal ganglia circuits are often termed cortical-striatal circuits. These circuits transfer information from the basal ganglia to the thalamus. Other subcortical structures closely connected to the basal ganglia are omitted in this sketch.

the basal ganglia, was coined centuries ago. The caudate nucleus and putamen, two contiguous structures, are the principal inputs to the basal ganglia. The putamen receives a stream of sensory information from other parts of the brain. It also monitors the completion of motor and cognitive acts. The caudate nucleus is active in a range of cognitive tasks. The globus pallidus, which is contiguous to the putamen, is the output structure for information. The information stream then is channeled to the thalamus and other subcortical structures.

Figure 2.3 shows three circuits, omitting most of the details, including links to other subcortical structures and posterior cortical areas. It is based on Cummings's 1993 review of clinical studies. Subsequent studies have confirmed its general conclusions and explored the neural bases of the behavioral deficits that Cummings reported. Cummings took into account the behavioral deficits of patients who had suffered strokes, comas, trauma, and Parkinson disease; the neural structures that were damaged or dysfunctional; and neural circuits mapped out by tracer studies of monkeys and other animals (for example, Alexander, DeLong, and Strick, 1986).

Conventional tracer techniques used on animals first mapped out circuits that connect various parts of the brain. These highly invasive techniques involve first injecting chemicals or viruses that form a "tag" into a specific neural structure. The tag essentially attaches itself to the electrochemical process by which neurons communicate with each other. Different tracer tags can propagate up or down a circuit's pathways. If the objective is to trace the neural circuits that control tongue movements, a "back-propagating" tracer tag could be injected into the tongue to trace the circuit backward. A "forward-propagating" tracer tag injected into a neural structure could determine whether it was part of a circuit that ultimately controlled tongue movements. After an animal was allowed to live for a week or so to move the tracer through the neural circuit, it was "sacrificed," that is, killed. The excised brain tissue then was stained with color couplers, similar to those used in conventional color film, that became attached to the tracer tag that had been transported through the neural circuit. The stained brain tissue, which showed the circuit in color, was visible after it was sliced in an apparatus that

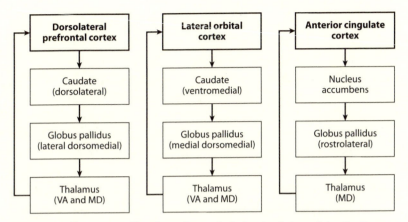

Figure 2.3. Three cortical-basal ganglia-cortical circuits that regulate attention, motor control, cognition, and emotion in human beings (omitting many details, including links to other subcortical structures and posterior cortical areas). The diagrams are based on Cummings's 1993 review of clinical studies. Subsequent studies confirm its general conclusions. Cummings took into account the behavioral deficits of patients who had suffered strokes, comas, trauma, and Parkinson disease; the neural structures that were damaged or dysfunctional; and the neural circuits mapped out by tracer studies of monkeys and other animals. The circuit from the anterior cingulate cortex involves the nucleus accumbens, which is also affected by alcohol and other substances.

produced very thin sections and then viewed microscopically. Another invasive technique directly monitors brain activity by driving "microelectrodes"—exceedingly minute electrodes—into an animal's brain to pick up the electrical signals that transmit information between neurons. Studies using multiple microelectrodes generally confirm the pathways mapped out by tracer studies (for example, Miller and Wilson, 2008). These techniques clearly cannot be used to map out neural circuits in human beings. Fortunately, noninvasive DTI confirms that humans have similar neural cortical-basal ganglia-cortical circuits (Lehericy et al., 2004).

The arrows connecting the labeled boxes in figure 2.3 track the flow of information from dorsolateral prefrontal cortex, lateral orbofrontal cortex, and ACC. The anterior cingulate circuit, as chapter 1 noted, regulates attention as well as laryngeal phonation (the melody of speech). Studies of the effects of brain damage from strokes and trauma showed that damage anywhere along the ACC circuit could result in a patient's becoming apathetic and/or mute. Virtually every neuroimaging (PET or fMRI) study that has ever been published shows ACC activity while subjects are paying attention to the task. The dorsolateral circuit is involved in cognitive tasks. Damage or dysfunction anywhere along this circuit (cortical or subcortical) and other circuits involving other prefrontal cortical areas, particularly ventrolateral prefrontal cortex, results in cognitive deficits ranging from being unable to understand the meaning of a sentence, to insisting on continuing to climb upward to the summit of Mount Everest when a blizzard is clearly approaching. The orbofrontal prefrontal circuit is involved in emotion, inhibition, and other aspects of behavior that define a person's personality.

The neuroanatomical terminology shouldn't put you off. For the most part, it simply specifies the location of each cortical area. Dorsolateral prefrontal cortex refers to the upper part of the side of the cortex. Ventrolateral prefrontal cortex is located below dorsolateral prefrontal cortex. Orbofrontal refers to the prefrontal area immediately behind the eyes. Note that the same neural structure, the caudate, receives information that is channeled into the globus pallidus before connecting to the thalamus and then back to prefrontal cortex. Information is transferred from one group of neurons to another in these neural structures in anatomically independent subcircuits, though these subcircuits appear to be interconnected by a web of dendrites in the globus pallidus.

Local Operations of the Basal Ganglia

The basal ganglia, in themselves, are not the key to human language. They must work in concert with parts of the cortex that are involved with motor control, cognition, and emotional regulation. The local operations of the basal ganglia first became evident through the study and treatment of Parkinson disease. The immediate cause of Parkinson disease is degeneration of the substantia nigra, a subcortical structure that besides forming part of the basal ganglia circuitry also produces the neurotransmitter dopamine. Neurotransmitters are hormones produced in the brain that are necessary for particular neural structures to function. Dopamine is one of the neurotransmitters involved in running the basal ganglia, and dopamine depletion is generally the primary cause of the motor, cognitive, and emotional problems associated with Parkinson disease (Jellinger, 1990). Levadopa, a medication that augments the level of dopamine in a patient's brain, and dopamine "agonists," medications that intensify the effect of remaining dopamine, can ameliorate some of the deficits of Parkinson disease.

Parkinson disease was first identified by Dr. James Parkinson in *An Essay on the Shaking Palsy* in 1817, about 150 years before levodopa treatment came into general use. The disease's motor problems can include tremor, muscular rigidity, slow movements, and an inability to carry out actions that require executing a series of internally guided sequential acts—walking, tying shoelaces, writing using a keyboard—most any learned activity. Patients generally do much better when they have an "external" model that they can copy. For example, patients who have a difficult time walking often are much better at imitating someone walking. These motor control problems stem from the failure of the basal ganglia's "normal" operation as a "sequencing engine" that calls out cortically stored submovements and executes them. The term "pattern generator" often is used to describe the set of motor commands that constitute an element of a motor action, such as walking, picking up a cup, or talking.

A Primer on Speech Production

A brief explanation of how human speech is produced is in order at this point because some of the most striking effects of basal ganglia damage are apparent when speech is analyzed. Computer-implemented systems can provide detailed analyses of movement patterns that yield insights on basal ganglia function.

Talking is arguably the most complex aspect of motor control that everyone has to master, which explains why even subtle speech production deficits can serve as markers for basal ganglia dysfunction.

Playing the violin or performing as a tightrope walker or tennis star undoubtedly requires the same or more precise degree of control, but everyone talks. It takes at least ten years, however, for normal children to learn to talk at adult rates (Smith, 1978). In part, children sound like children because their speech production is slow and full of errors.

A useful starting point in understanding what's involved in talking is to think about the way that tinted sunglasses color the world. The tint of the sunglasses results from a "source" of light (electromagnetic energy) being "filtered" by the sunglasses' colored lenses. The dye in the sunglasses' lenses acts as a filter, allowing maximum light energy to pass through at specific frequencies that result in your seeing the world tinted yellow, green, or whatever lens color you selected. The sunglasses' lenses don't provide any light energy; the "source" of light energy is sunlight, which has equal energy across the range of electromagnetic frequencies that our visual system sees as white. If the sunglasses' lenses remove light energy at high electromagnetic frequencies, the result is a reddish tint. If energy is removed at low frequencies, the result is a bluish tint.

Pipe organs work in much the same manner. The note that you hear when the organist presses a key is the product of the organ pipe acoustically filtering the source of sound energy produced by the air flowing through a constriction at the end of the organ pipe. The high airflow through the constriction produces acoustic energy across a wide range of frequencies, yielding "white" noise analogous to white light. The organ pipe reduces the amount of sound energy that can pass through it at most frequencies. The frequency at which maximum acoustic energy passes through the organ pipe is perceived as the musical note.

Human speech is a bit more complex. The human larynx (often called the "voice-box") generates a periodic source of energy for vowels and consonants, like [m]. The larynx is a complex bit of anatomy that sits on top of the trachea, the "windpipe" that leads upward from your lungs. It acts as a valve that can interrupt the flow of air from or into your lungs. The "vocal cords," sometimes called "vocal folds," act as a valve that opens the air passage when you breathe.

When you talk, the vocal cords are moved closer together and are tensioned by the muscles of the larynx. The airflow out from the lungs then sets them into motion and they rapidly close and open, producing puffs of air that are the source of acoustic energy for "phonated" vowels and consonants. such as [m] and [v] of the words "ma" and "vat." The rate, that is, the "fundamental frequency" (F_0), at which the vocal cords open and close depends on the pressure of the air in your lungs, the tension placed on the vocal cords and the mass of the vocal cords.

The average F determines the "pitch" of a speaker's voice. The fundamental frequency of phonation for adult males, who have larger larynxes than most adult women, can vary from about 60 Hz to 200 Hz. Adult women and children can have fundamental frequencies ranging up to 500 Hz, sometimes higher. The unit "hertz" (Hz) simply refers to frequency, the number of events per second. The periodic, or almost periodic, phonation also has acoustic energy at the "harmonics" integral multiples of the fundamental frequency of phonation. For example, if the F_0 is 100 Hz, there will be acoustic energy at 200 Hz, 300 Hz, 400 Hz, 500 Hz, and so on. Depending on how abruptly the vocal cords open and close, the acoustic energy produced by phonation falls off at higher harmonics to a greater or lesser degree.

While the perceived pitch of a person's voice is closely related to F_0, vowels and consonants are largely determined by "formant frequencies"—the frequencies at which the airway above the larynx can let maximum acoustic energy through it by acting as an acoustic filter, similar to an organ pipe. The difference between an organ and human speech-producing anatomy is that we have one "pipe"— the airway above the larynx, usually termed the supralaryngeal vocal tract (SVT)—that continually changes its shape as we talk by moving our lips, tongue, jaw, soft palate (a flap that can close off our nose from the rest of the airway), and the vertical position of our larynx. The continually changing shape of the SVT changes its filtering characteristics, producing a continual change in the formant frequency pattern. We'll return to this critical aspect of human speech, which yields its signal advantage over other vocal signals—a data transmission rate that is about ten times faster.

Speech Deficits of Broca's Syndrome and Parkinson Disease

The computer-implemented speech analyses that Sheila Blumstein and her colleagues published in 1980 showed that the primary speech production deficit of Broca's aphasia was errors in sequencing lip or tongue movements with phonation in words that began with the "stop" consonants [b], [p], [d], [t], [g], and [k]. Patients suffering from Broca's aphasia would produce the word "bat" when they intended to say "pat." I was aware of Sheila's research because we had jointly written a textbook on speech physiology and speech analysis. We and our graduate students found similar speech motor control sequencing problems in Parkinson disease (Lieberman et al., 1992; Pickett, 1998: Hochstadt et al., 2006).

Eliza Doolittle's Speech Lesson

Eliza Doolittle, who appears in George Bernard Shaw's play *Pygmalion* and the movie *My Fair Lady*, is based on a real person. Henry Sweet, one of the leading British phoneticians of the nineteenth century, taught a Cockney girl to speak "proper" English. It is said that Sweet never received the honor of a knighthood because he presented the tutored, real-life, Eliza to Queen Victoria at a social event.

In the movie, Eliza is shown enunciating proper English phrases before a candle flame. The candle's wavering flame served as a "burst detector" to mark the burst—the puff of air that happened when she opened her lips at the onset of a stop consonant. A strong burst would deflect the candle flame to a greater degree than a weaker burst. The burst for [b] should be less than [p] when the "voice-onset-times" that differentiate these consonants are correct, and so Eliza practiced speaking before a lit candle. She perhaps might have been rehearsing producing a mellifluous short-lag [d] and [b] for her future life when she requested her maidservant to "Please draw the bath."

Stop consonants, as their name implies, are formed by closing off the SVT (stopping it up) and then abruptly releasing the closure. The stop consonants [b] and [p] of words "bat" and "pat" are formed by the speaker's lips closing off the SVT; the distinction between them rests on the time that elapses between the burst of sound that occurs when the lips open and the start of phonation. The interval between the burst and phonation for the short-lag [b] is less than 25 msec (1 msec = 1/1000 of a second). The interval for the long-lag [p]'s exceeds 25 msec. The larger [p] burst is a consequence of the long lag between the burst and phonation (cf. Lieberman and Blumstein, 1988). Leigh Lisker and Arthur Abramson in 1964 analyzed the stop consonants of languages around the word. They found that every language that they analyzed (sound spectrographs, not candles, were used) differentiated stop consonants by means of this time-lag, which they termed "voice onset time" (VOT). In English, VOT differences differentiate short-lag [b], [d], and [g] from long-lag [p], [t], and [k]. The stop consonants [t], [d], [g], and [k] (of the words "to," "do," "go," and "come") are formed by obstructing the VOT with the tongue. A third VOT category "prevoicing," which isn't used in English, involves starting phonation before the stop consonant's releaseburst. Spanish, for example, uses prevoicing for the stop consonant of words beginning with the letters "b" and "d."

The Woman Who Suddenly Spoke Irish

As I listened to the messages recorded on my telephone one day in 1997, I learned that a case of "foreign accent syndrome" had been identified in Warwick, a suburb of Providence. A woman in her 40s, patient CM, had suddenly started to speak with a strong Irish accent when she came out of a coma. She wasn't Irish, nor had she ever visited Ireland. An MRI of CM's brain showed extensive bilateral damage to the putamen and caudate nucleus, the structures of the basal ganglia "striatum," which receives inputs from both motor cortex and prefrontal cortex. The query was whether my lab was interested in analyzing CM's speech, given our ongoing studies of Parkinson disease patients. We studied CM's speech, as well as her linguistic and cognitive capacities. Whereas the dopamine decrements in Parkinson can have some effects on cortical function, the lesions in CM's brain isolated the effects of basal ganglia dysfunction.

Our computer analysis showed absolutely no trace of an Irish accent. Instead, she had extreme problems coordinating and sequencing the motor acts involved in talking. She couldn't regulate the air pressure in the lungs in synchrony with gestures that moved her tongue, lips, jaw, velum (which closes off the nose), and the muscles that control laryngeal phonation. VOT was extremely disrupted. Sudden changes in the amplitude of her speech randomly occurred; her voice would abruptly fall to a whisper and then become suddenly loud. Her vocal cords were phonating where they shouldn't. The speech therapists probably were at their wits' end and decided that she was speaking with an Irish accent because her speech was unlike any that they had previously heard.

Despite her ostensible foreign accent, CM could be understood. Her major problem was cognitive; her daily life was constrained by the fact that she couldn't plan ahead or adjust to unanticipated events, however minor. She needed overt cues; her refrigerator door had a daily schedule that listed the routine tasks that she had to do throughout the week. Similar problems beset Parkinson disease patients, who have difficulty with routine tasks that require executing a sequence of internally guided motor acts, or planning ahead. External cues make the task easier. The most severe cognitive problem, which CM shared with Parkinson disease patients, was cognitive perseveration—the inability to change the direction of a thought process.

Cognitive Flexibility—Perseveration

In the 1980s, it became apparent that Parkinson disease has negative cognitive consequences. Cognitive inflexibility, characterized as "subcortical dementia," occurred. Subcortical dementia is different in kind from Alzheimer's, which primarily affects memory. It shows up as perseveration—continuing to act or think along one line, though circumstances dictate a change. The Odd-Man-Out (OMO) test was devised by Flowers and Robinson (1985) to assess cognitive flexibility in Parkinson disease.

The test is straightforward. Healthy "control" subjects usually are puzzled why anyone would bother administering a test that is so simple. The person being tested is presented with a booklet that has two sets of 10 cards. Each card has three images printed on it. The first card might have a large triangle, a small triangle, and a large circle. The subject is asked to identify the "odd" image. There are two ways, "criteria," that the subject can use to reach that decision—the subject can use either shape or size to decide which image is "odd." If the criterion selected by the subject is size, the small circle will be selected. The second card also has three images printed on it, an uppercase *E*, a lowercase *e*, and an uppercase A. The subject is asked to select the odd image using the same criterion, and is told whether his/her decision is correct or not after each trial. The subject starts with the first 10-card set; after s/he comes to card 10, the subject is asked "to do the sort another way" for the next 10-card set. And after 10 trials, the subject is again asked to re-sort the same packet of cards "the other way." This entails returning to the criterion that s/he used to sort the first 10-card set, five minutes or so earlier. The procedure is repeated six or eight times. Figure 2.4 shows two Odd-Man-Out cards.

Control subjects generally have a few errors on the first 10-card sort and none thereafter. In contrast, Parkinson patients often make few or no errors on the first set of 10 cards, but they encounter prob-

Figure 2.4. Two cards from the Odd-Man-Out test.

lems on each shift of the sorting criterion—shifting from size to shape, or from shape to size. They even have high error rates when they return to the first 10-card set on which they made few or no errors. Thirty-percent error rates for subjects having basal ganglia damage typically occur on the OMO test. The problem rests in shifting the sorting criterion; Parkinson disease patients perseverate—they keep to the criterion that they have previously been using.

When CM, the woman who spoke Irish, was tested by Emily Pickett and her colleagues, she was at a complete loss whenever she was asked to change the sorting criterion. She stared helplessly at the test cards, incapable of changing the sorting criterion (Pickett et al., 1998). CM was not demented; she understood what she was supposed to do and knew that she was not able to perform this simple task. She repeatedly asked for explicit guidance. CM also had problems on a range of cognitive tasks that involve "executive control," such as remembering images that she had viewed a few minutes earlier, and suppressing a response on the Stroop Test, in which colored inks are used to spell out color terms. When asked to identify the color of red or green ink that spelled out the word "blue," she had difficulties. She also had high error rates on a sentence comprehension test on which children older than age eight years have no problems. Her error rates increased dramatically for sentences in which the syntax departed from "canonical" sentences that had no clauses, for example, "The girl kissed the boy." CM had difficulty deciding who kissed whom when she heard the similar sentence in the passive voice, "The boy was kissed by the girl." Problems comprehending distinctions in meaning conveyed by syntax also beset Parkinson disease patients (Lieberman et al., 1990, 1992; Grossman et al., 1992, 1993, 2001; Hochstadt et al., 2006).

Motor Control and Cognitive Flexibility

The speech motor sequencing problems apparent in Parkinson disease and instances of damage confined to the basal ganglia in experiments-in-nature such as the study of CM have a common cause, a degraded basal ganglia local operation. In 1994, the British neurosurgeons

David Marsden and Jose Obeso, in their comprehensive review of the effects of neurosurgery aimed at ameliorating the motor control problems of Parkinson disease, identified two aspects of basal ganglia motor control. The first basal ganglia motor control function explains why Parkinson disease patients and Broca's aphasics can have difficulty sequencing the exceedingly precise motor commands that are necessary to produce stop consonants:

> First, their [the basal ganglia] normal routine activity may promote automatic execution of routine movement by facilitating the desired cortically driven movements and suppressing unwanted muscular activity. Secondly, they may be called into play to interrupt or alter such ongoing action in novel circumstances (Marsden and Obeso, 1994, p. 889).

Marsden and Obeso were aware of the cognitive deficits that can occur in Parkinson disease, and the dual roles of the basal ganglia in motor control led them to conclude that:

> The basal ganglia are an elaborate machine, within the overall frontal lobe distributed system, that allow routine thought and action, but which respond to new circumstances to allow a change in direction of ideas and movement (Marsden and Obeso, 1994, p. 893).

As we shall see, this insight, which explained the cognitive inflexibility that occurs in Parkinson disease, is a key to understanding the neural basis and probable evolution of human cognitive flexibility and creativity.

Our Laboratory at Mount Everest

For almost 10 years, Mount Everest served as the site for an experiment-in-nature that further explored the role of cortical-to-basal ganglia circuits in motor control, language, and cognitive flexibility, as well as keeping one's impulses under control (Lieberman et al., 1995, 2005; Lieberman, 2006). The basal ganglia are metabolically active and need a copious supply of oxygen to function properly. Basal ganglia function thus is degraded by the low oxygen content of air at the extreme altitudes reached on Himalayan peaks. The effects generally do not show up at the lower altitudes of the Alps and most other mountain ranges. Climbers ascending Everest and other 8,000-meter-high Himalayan peaks have exhibited Parkinson-like motor deficits and subcortical dementia. In some instances, they have collapsed and died. Autopsies and MRIs in these instances reveal bilateral lesions in the putamen and caudate nucleus (Chie et al., 2004; Swaminath et al., 2006). Cognitive deficits, similar to those seen in Parkinson disease, occur when climbers ascending Mount Everest are seemingly doing fine, but extreme motor and cognitive deficits resulting from basal ganglia dysfunction contributed to the death of one climber (Lieberman et al., 2005; Lieberman, 2006).

Only a handful of climbers have ever reached Everest's summit in one long, sustained climb. The ascent almost always involves climbing up to a high "camp" and then descending to base camp, resting, climbing up to a higher camp and descending. After repeating this sequence and reaching successively higher camps over a period of weeks, climbers ultimately reach the highest camp at 8,000 meters before attempting to make a final push to the 8,848-meter-high summit. This allowed my research team to run a series of motor and cognitive tests on the same individual as he or she reached higher and higher altitudes and breathed "thinner air" that had lower and lower oxygen levels. We remained, nerd-like, at base camp and used radio links to administer linguistic and cognitive tests and record the climbers' speech.

Climbing Everest also differs radically in another way from the manner in which mountains are usually climbed. Without the aid of the Sherpas, a Tibetan people who moved to the Everest region of Nepal 300 years ago, almost no one else would be able to reach Everest's summit. Before any other climbers leave base camp, Sherpas string miles of "fixed" rope. A rope line is anchored to the slopes between camp 2 at 6,500 meters and camp 4 at 8,000 meters. Climbers then ascend and descend with the aid of the fixed rope line using clamping devices attached to their safety harnesses. As they push onward and upward on the rope line, the brightly colored down-filled garments of the long line of climbers on summit days (the few days on which the wind dies down so that they can reach the summit) from a distance look like a long red and yellow caterpillar moving upward.

We found correlated speech production deficits, deficits in comprehending the meaning of sentences that eight-year-old children can understand, and perseverative errors on the Odd-Man-Out test and Wisconsin Card Sorting Test (described later), which provides another measure of cognitive flexibility. The errors on these tests and degraded speech effects were similar to those evident in Parkinson disease and subjects like speaker CM. The climbers' speech became slower, and error rates on these tests increased. We could predict cognitive and sentence deficits by means of computer-implemented acoustic analysis of the climber's speech.

The effects were subtle in most instances. However, profound cognitive and motor dysfunction had a tragic outcome in one instance. At Everest base camp, the 5,300-meter altitude starting point, one 23-year-old male climber was error-free on the Odd-Man-Out test. His speech was normal when analyzed using a computer system that reveals subtle deficits that are not apparent to the unaided ear. Two days later, when he was tested by two-way radio shortly after he reached camp 2 at 6,500 meters, his performance had degenerated to a pathologic level. His OMO error rate exceeded 40 percent. The computer analysis revealed profound speech motor sequencing deficits similar to the brain-damaged woman CM studied in Pickett et al. (1998). The climber was forcefully advised to descend. Tragically, his inability to change the direction of a thought process was not limited to the OMO test. His cognitive perseveration was general, and he kept to his original climbing plan, which called for him to ascend to camp 3 at 7,200 meters. Once there, he attempted to return to base camp. Motor sequencing problems then contributed to his death. Two "carabiners," snap links tethered to each climber's safety harness, link each climber to the fixed ropes. The carabiners' function is to prevent the climber from sliding down the icy slopes. Two carabiners are necessary because the fixed rope is anchored into the slope every 5 meters or so by snow-stakes or ice screws. It's necessary to first unsnap the downhill carabiner, move it past the past the anchor point and refasten it to the

47

rope. The uphill carabiner must remain in place above the anchor while the downhill carabiner is unlocked and moved. The uphill carabiner then can be moved past the anchor point after the downhill carabiner is refastened to the fixed rope. At some point during his descent, the oxygen-starved 23-year-old climber failed to carry out the correct carabiner sequence and fell to his death.

Cognitive inflexibility induced by low oxygen levels accounted for the disaster on Everest described in Jon Krakauer's book, *Into Thin Air*. The expedition leaders, who had years of Himalayan experience, "perseverated," keeping to their original climbing plan toward Everest's summit though a major storm was approaching. They should instead have immediately descended. Similar cognitive inflexibility is thought to be the immediate cause of aviation disasters and near-disasters over the past century. Engines run out of fuel and planes crash because pilots fail to switch to full auxiliary fuel tanks. The 2008 PBS documentary *Redtail Reborn* about the World War II African-American Tuskegee fighter group re-created a near-fatal incident. The oxygen-mask of one pilot malfunctioned. He lost control and his plane went into a steep dive while he hallucinated, talking to a stranger who somehow was sitting astride the plane's nose. He survived because the cockpit canopy was open and the oxygen-rich outside air at a lower altitude revived him before the plane hit the ground. Climbers on Mount Everest suffering from hypoxia also have reported conversing with hallucinatory companions.

Gage Is No Longer Gage — Disinhibition

Another effect of oxygen deficits became evident during the nine years that we tracked motor and cognitive performance in climbers on Mount Everest. Some climbers became disinhibited, losing control over impulsive behavior. This tended to occur in climbers who had spent lots of time at altitudes above 6,500 meters, or had attempted the climb without ever using supplementary oxygen. Disinhibition on Everest from hypoxic disruption of basal ganglia activity approaching clinical levels on Everest wasn't surprising. The story of Phineas

Gage is a staple of neuroscience texts. Gage was a nineteenth-century railway worker who survived after a three-foot-long iron rod, propelled by a blasting charge, passed through his brain. Gage's miraculous survival was marred by a personality change—disinhibition. The iron rod destroyed Gage's orbofrontal prefrontal cortex. Dr. Harlow, the physician who attended Gage, noted the change,

> He is fitful, irreverent, indulging at times in the grossest profanity (which was not previously his custom), manifesting but little deference for his fellows, impatient of restraint or advice when it conflicts with his desires, at times pertinaciously obstinate. . . . Previous to his injury, although untrained in the schools, he possessed a well-balanced mind, and was looked upon by those who knew him as a shrewd, smart businessman, very energetic and persistent in executing all his plans of operation. In this regard his mind was radically changed, so decidedly that his friends and acquaintances said he was "no longer Gage" (Harlow, 1868, p. 342).

The disinhibited Everest climbers we observed acted impulsively and talked in a manner that approached the clinical cases reviewed by Cummings (1993) for disruption to circuits linking the basal ganglia and orbofrontal prefrontal cortex.

Executive Control

The mechanisms involved in treating Parkinson disease using levadopa are not fully understood. After granting a temporary levadopa "holiday" to the Parkinson disease patient, it often becomes ineffective. When it works, many Parkinson disease patients respond well over the course of an entire day. In other "fluctuating" patients, the effects of medication rapidly wear off. In some circumstances, patients are taken off levadopa to assess the state of Parkinson disease, unmasked by medication. This sensitivity to medication creates an opportunity for exploring the effects of basal ganglia dysfunction, since cortical function, for the most part, depends on different neurotransmitters. The same person, "on" or "off" medication, can function as his or her own "control" in experiments that systematically test motor and cognitive abilities. Trevor Robbin's research group at Cambridge University in England, in a series of elegant experiments (for example, Lange et al., 1992), showed that medicated Parkinson disease patients can perform at levels close to unimpaired normal controls on a range of cognitive "executive control" tasks. As chapter 1 noted, executive control encompasses tasks such as problem solving, decision-making, visual tracking, and tasks involving temporary, short-term, working memory as well as cognitive flexibility.

Working memory refers to the ability to hold and act on information that is not directly present in the environment. In experiments that date back to the 1920s, monkeys were first trained to push a button positioned at a specific location on a control board to obtain food. They could look at the button during the training period. When the control board was obscured, the monkeys then were for a short period still able to obtain their food reward by poking at the same position on a blank panel. Baddeley and Hitch (1974) pointed out that many of the tasks that constitute human linguistic and cognitive capabilities involve working memory. Mentally adding or subtracting numbers, for example, involves keeping at least one number in working memory. Sorting out images according to a desired shape or color involves retaining the desired "target" shape or color in memory. Comprehending the meaning of a spoken sentence involves keeping the words in working memory because the sentence's meaning is not clear unless you take account of all the words. Baddeley and Hitch posited an "executive" mental capacity that directs the cognitive acts involved in performing these tasks. Hence the term "executive control" came into being to refer to these tasks.

It was first thought that the short-term "working memory" that enabled the monkeys to perform this task involved temporarily storing information in prefrontal cortex, because neurons there were active during the time interval in which a monkey could perform the task. Current studies suggest that working memory does not actually entail storing information in prefrontal cortex. The details of how this selection process works and what cortical structures are involved are still a work in progress, but data from many neuroimaging studies (for example, Postle, 2006; Badre and Wagner, 2006; Miller and Wallis, 2009) indicate that ventrolateral, dorsolateral, and other prefrontal cortical areas in neural circuits involving the basal ganglia direct attention to information stored in the brain systems that directly learn, store, and execute concepts and actions. The information then is used to carry out tasks involving executive control, such as comprehending the meaning of a sentence (Kotz et al., 2003), mental arithmetic (Wang et al., 2006), and selecting words according to their meaning or sound structure (Simard et al., 2011). My laboratory, working with Dr. Joseph Friedman, a Parkinson disease specialist, showed that Parkinson disease patients had difficulty comprehending sentences where distinctions in meaning conveyed by syntax involve holding words in working memory (Lieberman et al., 1990, 1992; Hochstadt et al., 2006). The findings have replicated in other laboratories (for example, Grossman et al., 2001, 2002; Natsopoulos, et al. 1993).

Neuroimaging

Although there sometimes is a tendency toward reductionist thinking in some fMRI and PET neuroimaging experiments (locating the "center" of morality or syntax, and so on), the vast majority are advancing our understanding of how human brains work. Oury Monchi's research group in Montreal, which is one of the world's centers of research on the brain, has separated out some of the critical functions performed by the structures that form the basal ganglia. The Monchi research group in a sequence of fMRI imaging experiments has teased apart some of the local operations that yield human cognitive flexibility using the Wisconsin Card Sorting Test (WCST), which is the gold standard for assessing cognitive flexibility. The conventional WCST uses a set of 18 cards and 4 "reference" cards. Each test card has pictures of one of four shapes—circles, stars, squares, and plus signs (+)—printed in one of four colors. There can be one, two, three, or four circles, stars, squares, or plus signs on a card. The subjects have to match each test card to one of the four reference cards, for example, a card with one red triangle, a card with two green stars, a card with three yellow plus signs, a card with four green circles. A typical test card might have four yellow stars printed on it. The subject isn't told how she or he should sort the test cards. The subject starts out by making a sort and then is informed whether the sort was "correct" or not.

For example, starting with number and matching the test card to the reference card that has four green circles on it would be incorrect if the person running the session wanted to start with color. The subject in this instance would be told that the sort was "wrong" and have to continue to make sorts until s/he matched to the color on one of the four reference cards, when s/he would be told that the sort was "correct." The subject then would continue to match by color, receiving a "correct" response until the sorting criterion was again changed by the test administrator. After achieving a correct response to the new criterion, for example, number, the subject would then receive "positive feedback" — being informed that the sort was correct — and would continue until the sorting criterion changed again. The WCST has been in use since the 1960s to study the cognitive role of frontal cortical activity in humans and shows whether cognitive flexibility is degraded when subjects have difficulty changing the sorting criterion, or learning a new sorting criterion, as well as maintaining a cognitive criterion. The WCST also can be used to determine the specific roles of of basal ganglia structures in these tasks.

The Monchi et al. (2001) fMRI study reported the involvement of a cortical-basal ganglia loop involving the ventrolateral prefrontal cortex, the caudate nucleus, and the thalamus when a subject received "negative" feedback, indicating the need to plan a cognitive set-shift by looking for an alternative rule-sorting criterion. Another cortical-striatal loop, including the posterior prefrontal cortex and the putamen, was active during the execution of a set-shift — that is, when applying the new sorting criterion for the first time. Dorsolateral prefrontal cortex was involved whenever subjects received any feedback, positive or negative, as they performed card sorts, to apparently monitor whether their responses were consistent with the chosen criterion. In another fMRI study, Monchi et al. (2006a) showed that ventrolateral prefrontal cortex is specifically active in planning ahead when a subject has to compare the card that is to be selected and the criterion. This study confirmed that the caudate nucleus uses this information when any novel action needs to be planned.

The role of the caudate nucleus, putamen, and the subthalamic nucleus, and other subcortical structures involved in these circuits (not shown in the simplified 1993 Cummings diagram; cf. Alexander et al. for a fuller picture) became clearer in this study and others by the same group (Monchi et al., 2006a, 2006b; François-Brosseau et al., 2009; Provost et al., 2011). The results of these studies point to the caudate nucleus being involved in the selection and planning, and the putamen in the execution of a self-generated action among competitive alternatives. Neither was active when the information for shifting to an alternative was implicitly given in the task. In contrast, the subthalamic nucleus (another basal ganglia structure) was involved only when a new motor action was required, whether planned or not.

Other neuroimaging studies confirm the role of dopamine and the basal ganglia in cognitive flexibility. When early-stage Parkinson patients who have low dopamine levels were compared with healthy control subjects in another study, a pattern of reduced activity in both dorsolateral prefrontal cortex and left ventrolateral prefrontal cortex was apparent in "executive control" tasks (Monchi et al., 2007). Parts of parietal and temporal cortex, posterior parts of the brain that these cortical-basal ganglia circuits connect to prefrontal cortex, also showed reduced activity compared to healthy subjects when the putamen or caudate nucleus of the basal ganglia were required for the task.

Cortical-Basal Ganglia Circuits Are Not Domain-Specific

As noted earlier, the local operations performed in a particular neural structure aren't necessarily domain-specific—devoted to one task and one task alone. That is clearly the case for the basal ganglia. Independent neuroimaging studies have reached similar conclusions concerning the role of specific areas of the prefrontal cortex and basal ganglia in a range of cognitive tasks. Cools et al. (2008), using PET imaging, showed that working memory tasks involving recalling digits, words, images, reading, and so on all correlated with basal ganglia dopamine levels. Hazy et al. (2006) showed that cortical-basal ganglia circuits integrate activity in prefrontal cortex, basal ganglia, and the subcortical hippocampus in these tasks, which accounts for the higher basal ganglia dopamine levels that occurred in these tasks. Wang et al. (2005) found increased activity in left ventrolateral prefrontal cortex and the basal ganglia as task difficulty increased during a mental arithmetic task. Kotz et al. (2003), using event-related fMRI, found bilateral ventrolateral prefrontal cortical and basal ganglia activity in tasks that involved either comprehending the meaning of a spoken sentence or the emotion conveyed by the speaker's voice. These studies confirm the conclusions of Duncan and Owen, who reviewed the findings of neuroimaging papers published before 2000. They concluded that mid-ventrolateral and dorsolateral prefrontal cortex, as well as the anterior cingulate cortex, are active in virtually every cognitive task involving executive control. There may be some neural structures devoted strictly to language, but that is becoming less and less likely.

It also is becoming evident that the cortical to basal ganglia circuits that we have been discussing are not specific to cognition, insofar as they are involved in regulating emotion. But are there specific circuits, constituting "modules," that are domain-specific. fMRI studies using variations on the Wisconsin Card Sorting Test suggest that the observed neural activity in the cortical-basal ganglia circuits that carry out this task is not domain-specific—limited to sorting out visual symbols. France Simard and her colleagues in their 2011 study had subjects match words according to their meaning (semantics), their phonetic similarity (roughly their spelling), and whether they rhymed (the syllable's onset sounds). The same activation patterns occurred (involving ventrolateral prefrontal cortex, dorsolateral prefrontal cortex, caudate nucleus, and putamen) as was observed when sorting images. Additional cortical areas involved with sound perception also were activated. The critical point was an identical pattern of prefrontal cortex to basal ganglia circuits activity while shifting cognitive criteria in sorting tasks involving either visual or linguistic criteria.

The local operations performed in the thalamus, which has many distinct areas, are not clear, but it has been apparent for decades that lesions in thalamus, particularly in areas that form part of the circuits to the basal ganglia, can result in speech motor, word-finding, and comprehension deficits (Mateer and Ojemann, 1983). The surgical procedures used to treat Parkinson disease before levodopa was available entailed producing deliberate unilateral lesions in the basal ganglia circuitry that projects to the thalamus or bilaterally. Similar surgical procedures directed by MRIs and electronic stimulators implanted in basal ganglia, thalamus, and other nearby subcortical structures are again being used to tread Parkinson disease. In some instances, especially with bilateral lesions of the globus pallidus (Scott et al., 2002), cognitive deficits occur. In some instances, no cognitive deficits are apparent. I saw a film of a young boy who had been unable to stand upright or walk, running after surgery that caused bilateral lesions in globus pallidus; no cognitive deficits were apparent. As Marsden and Obeso (1994) pointed out, our knowledge of neural circuitry and local neural operations remains imperfect.

So You Want to Be a Samurai, Play the Piano, or Understand German

The basal ganglia have yet another role that isn't domain-specific — associative learning and "automatization." Many of the tasks that we carry out throughout life involve the formation of "matrisomes," groups of neurons in motor cortex that control the muscles that carry out a learned task rapidly, without conscious thought. The data of an fMRI study aimed at showing microcortical areas in individual subjects specific to understanding sentences and words (the experimental protocol didn't achieve that end) suggests that a similar process forms "cognitive" and "linguistic" matrisomes in prefrontal cortex (Fedorenko et al., 2011).

One of the movies that spoke to my wife and me in the late 1970s was the Japanese film *Yojimbo*. We were in the midst of a legal battle against the overwhelming resources of the University of Connecticut. Marcia was on the front line of the feminist struggle for equality. She had been denied tenure because she had pushed for equal access to UConn's athletic facilities for women. In *Yojimbo*, Toshiro Mifune, in his role of a wandering Samurai, defeated a villainous band. We needed a Yojimbo. We viewed *The Hidden Fortress*, *The Throne of Blood*, and other Japanese slice-'em-ups. The precise, blazingly fast Samurai sword cuts were incredible. How did the Samurai do it?

The path to mastering the Samurai sword was laid out in 1645 by Miyamoto Musashi in *The Book of the Five Rings*. Musashi's lesson plan applies with equal force to talking, playing the piano, touch-typing, and the complex protocols that mark human culture, including learning the linguistic "rules" of syntax and word-formation. If you've successfully taught someone how to drive a car with a stick shift, you've probably followed Musashi's formula. To master the Samurai sword, or gearshift and clutch, you must slowly perform the action, repeating it again and again. At first, you'll have to consciously think about each step. But at some point, you'll suddenly realize that you are not consciously aware of the steps involved and can go on to thinking about other things. Research started in the 1970s shows that many neural structures besides motor cortex are active while you are engaged in learning the task. However, when the task becomes automatic, most of the neural structures that apparently were thrown into action to learn the task are no longer active. They constitute a sort of "temp-help" that facilitate learning the task; they're no longer needed (cf. Sanes et al., 1999). And at the point where an action becomes automatic, "matrisomes," populations of neurons, are formed in motor cortex that code the entire action. They code the submovements, the motor acts necessary to carry out the task automatically. Marsden and Obeso (1994) correctly pointed out that the basal ganglia sequentially call out the submovements stored in motor cortex when we walk, talk, or perform any motor act.

Learning Anything

Studies of motor learning have revealed another local basal ganglia operation—facilitating a reward-based associative learning process—

making it even clearer that the local operations of the basal ganglia are not domain-specific. Mirenowicz and Schultz (1996) monitored basal ganglia dopamine-activated neurons in the basal ganglia while monkeys learned how to obtain a juice reward by pushing a button. The neurons were activated when the monkeys were rewarded. When MPTP, a neurotoxin that degrades the dopaminergic basal ganglia activity, was administered, the monkeys were unable to obtain their juice reward. When the monkeys were given dopamine "agonists," used to treat Parkinson disease by restoring basal ganglia activity, the monkeys again pressed the buttons to get the juice. Using somewhat different procedures, Graybiel (1995) replicated these findings. Her studies show that the basal ganglia play an essential role in learning both motor and cognitive acts.

The neural network that makes up the basal ganglia involves two integrated subsystems. A "fast" system that involves the neurotransmitters glutamate and GABA connects the cortex with the basal ganglia's input—the striatum, which includes the putamen, caudate nucleus, and the basal ganglia's output structures. A second subsystem arising from the midbrain makes use of the neurotransmitters dopamine and acetylcholine in "interneurons" that connect neurons within the basal ganglia (Bar-Gad and Bergman, 2001). The neurons of these subsystems are among the elements that are the brain's memory traces of the relationships formed through associative learning. In an experiment that used classic Pavlovian techniques, Mati Joshua and colleagues (2008) studied neuronal basal ganglia activity in monkeys after they learned to perform tasks that resulted in either a "reward" (fruit juice) or an "aversive stimulus" (a puff of air directed on their face). The experimental technique involved placing microelectrodes into the monkeys' brains. Increased activity in dopaminergic basal neurons coded the probability that a reward will be achieved. This explains the results of degraded dopamine activity noted by Mirenowicz and Schultz as well as clinical observations of Parkinson disease patients, who show less facility in learning tasks that would result in a pleasing outcome. In contrast, increased activity occurs in the monkey's interneurons when an aversive outcome is anticipated (Joshua et al., 2008). Wael Assad and Emad Eskandar (2011) monitored neurons in monkeys as they performed trial-and-error learning. They found neurons that coded the monkey's expectations of potential reward (juice) or aversive puffs of air in both the caudate nucleus of the basal ganglia and the prefrontal cortical areas that are linked to the caudate. These coding processes allow monkeys to learn to perform long sequences of actions that have "selectional constraints." At each step of the sequence, the monkey must learn to perform a particular act and choose not to perform other acts if he is to achieve his goal (a food reward). The monkey thereby learns to perform a series of acts that if we were to use linguistic terminology, have an action syntax.

In short, these neurons code the expectations that guide associative learning, allowing animals to learn to perform complex linked sequences. In humans, similar processes would account for our learning complex grammatical "rules," as well as the equally complex rules that guide our interactions with other people, other species, the conditions of daily life, and responses to new and novel situations.

Evie Fedorenko and her colleagues at MIT in their 2011 fMRI study attempted to identify locations in Broca's area that they believe are innate and specific to comprehending sentences and words. However, these micro-areas probably are learned matrisomes that code the suite of neural operations that are necessary to comprehend a class of sentences. Automatized motor acts are controlled by matrisomes formed in motor cortex. Ongoing projects developing brain-controlled prostheses and speech-producing synthesizers for individuals who have lost limbs or have suffered profound muscle degeneration are using electrodes to tap into these matrisomes. There is no reason to believe that other cortical areas, including prefrontal cortex, engage any neural processes that differ from those carried out in motor cortex.

The studies noted earlier (a small subset of current studies) show that the outcomes of tasks learned by means of associative learning are coded by neurons in the basal ganglia and prefrontal cortical areas linked to basal ganglia. The pervasive presence of cortical malleability suggests that cognitive matrisomes that code the "rules" by which we comprehend sentences and the operations involved in mathematics, formal logic, and so on are also formed by means of associative learning, imitation, and other "general" cognitive processes. One of my students may hold the current world's record for being the subject of fMRI studies. She is totally blind and "reads" using text-to-speech software on her computer. Her "visual" cortex responds to speech and other vocal signals; she has been the focus of fMRI studies directed at mapping out the cortical activation that enables her to recognize spoken words.

It is probable that we form cortical cognitive matrisomes, populations of neurons that code the semantic referents of words and syntax of a language as we learn it. The cognitive matrisomes would be formed by the process of automatization that we know takes place in motor cortex as we learn to tie our shoes, shift gears, or for the students of Toshiro Mifune, mastering the Samurai sword.

But, as Miyamoto Musashi observed, automatization takes time. Learning to tie your shoes was a slow process. Learning to talk takes years; normal children don't achieve the same speaking rate as adults until age 10 (Smith, 1978). The same time frame is necessary to learn syntax, contrary to the claim of linguists, who have adopted the theoretical framework proposed by Noam Chomsky. We will return to the question of the "faculty of language" in the chapters that follow.

Chapter Three

Darwin Got It Right

Charles Darwin's theory of evolution was attacked by virtually all of his academic colleagues when he published *On the Origin of Species* in 1859. Darwin's theory still remains controversial. When surveyed, about 40 percent of Americans and 25 percent of Britons still say that they don't "believe" in evolution. The word "believe" is significant because a statement affirming belief may be relevant if the question concerns the Virgin Birth or the Tibetan deity Mahakala, but accepting any scientific theory—gravity, evolution, global warming— rests on whether it correctly predicts observable events. But Darwin would have been amazed, perhaps dismayed, by the continuing flood of scholarly books and papers that attempt to show that he was wrong.

However, it's first necessary to understand what Darwin actually proposed. Although Darwin was unaware of genes or DNA and obviously lacked access to the overwhelming body of evidence that has since supported his theory, he got it right. The basic premises and research paradigms of evolutionary biology, stated in a clear manner in the first edition of *On the Origin of Species*, are still valid. You, yourself, can judge whether this is the case because the *Origin* is free of jargon; biological jargon hadn't been invented. I've used a facsimile edition for many years in my courses.

Darwin also raised issues that transcend the debates that usually occupy academics, issues that are relevant today, such as family planning and the impact of human activity on the environment. He showed that seemingly neutral changes driven by political and economic forces can have profound consequences on the environment. English villages and towns, for example, once had "commons," the green open spaces, often ringed by eighteenth- and nineteenth-century homes that grace New England towns. The common was a place where anyone could graze livestock, but in the early years of the nineteenth century a land-grab occurred. The commons were privatized and "enclosed." The ecosystems within the walled and fenced enclosures quickly changed. Absent grazing livestock, plants, shrubs, and trees sprung up. In turn, the insects, birds, and animals living within the enclosures totally changed.

The finches that Darwin had collected and observed in the Galapagos Islands during his voyage on the *Beagle* played a central role in Darwin's thoughts on the "transmutation" of species. The 13 different Galapagos finch species have different food sources, insects living in the ground, cactus seeds, and different types of nuts that determine the shapes of their beaks. Darwin became convinced that the different species had each been transmuted through a series of small changes to best exploit particular ecosystems. It has since been shown that changes in rainfall rapidly change the selective forces acting on the birds. For example, dry weather favors the survival of finches that have short, strong beaks capable of splitting open hard-shelled seeds; wet weather favors long thin beaks. Changes in the ecosystem, shifting the selective forces on the bird population, in turn affect the birds' behavior and bird calls (Podos, 2001).

The Tenuousness of Life and Natural Selection

Darwin also was aware of the earlier evolutionary theories of his grandfather, Erasmus Darwin, and John-Baptiste Lamarck. On his return to England from his voyage on the *Beagle* in 1835, Darwin was wondering how the fossils of extinct animals that he had found in South America could fit with theories that posited the immutable nature of species. However, the missing element in previous evolutionary theories was a directing mechanism that did not involve divine direction. Darwin's notebooks show that he grappled with this problem for years. The insight that provided the solution derived from a problem that is still with us—human population growth. Europe's population had been almost static since the sixteenth century until it began to increase during the eighteenth century and doubled by 1900.

Romantics may yearn for the "good-old-days" in preindustrial Europe. Life was pleasant for the privileged few, but as Robert Fogel points out in his book *The Escape from Hunger and Premature Death, 1700–2100*, life was miserable for almost everyone else. In the early years of the nineteenth century:

> The supply of food was not only meager in amount, but also relatively poor in quality. . . . Even prime-age males had only a meager amount of energy available for work (Fogel, 2004, p. 9).

Into this ecosystem came the first fruits of public health. We can all be thankful that virtually everyone ignored Victor Hugo's agricultural advice. In his novel, *Les Misérables*, in a discourse set apart from the story of Inspector Javert's relentless pursuit of Jean Valjean, Hugo lamented the fact that farmers were no longer using "night soil" (human waste) to fertilize their crops.

The segregation of human fecal material from the food and water supply was introduced on a broad scale in Europe. Communicable diseases decreased, and an extraordinary increase occurred in lifespan and population size. In 1750, 75 percent of all children born in London were dead before their fifth birthday. Life expectancy in England was 37 years. Life expectancy in France was only 26 years. But over the next fifty years, sanitation improved and Europe's population doubled. By the year 1810, 78 percent of all children born in London lived past their fifth birthday. The population of Europe was increasing at an unprecedented rate. Thomas Malthus's 1798 tract, *An Essay on the Principle of Population*, predicted incipient disaster.

Malthus thought that the population surge was outstripping the food supply. Famine, pestilence, disorder, and war would result.

The impending disaster never happened because agricultural production outpaced population growth, but Malthus's "essay" led to Charles Darwin's "Eureka" moment. On reading the essay, Darwin realized that the

> doctrine of Malthus applied with manifold force to the whole animal and vegetable kingdoms, for in this case there can be no artificial increase of food, and no prudent restraint from marriage . . . and a struggle for existence inevitably follows from the high rate at which all organic beings tend to increase (Darwin, 1859, p. 63).

Darwin's terminology, the "struggle for existence," probably led to Alfred Lord Tennyson's view of "Nature red in tooth and claw," and to films on evolution beginning with scenes of lions devouring their prey. However, Darwin stated instead that

> I use the term Struggle for Existence in a large and metaphorical sense, including dependence of one being on another, and including (which is more important) not only the life of the individual, but success in leaving progeny (Darwin, 1859, p. 62).

Natural selection, the key element of Darwin's theory of evolution, followed from this insight. As Darwin put it,

> any variation, however slight and from whatever cause proceeding, if it be in any degree profitable to an individual of any species, in its infinitely complex relations to other organic beings and to external nature, will tend to the preservation of that individual, and will generally be inherited by its offspring. The offspring also will thus have a better chance of surviving, for, of the many individuals of any species which are periodically born, but a few can survive. I called this principle, by which each slight variation, if useful, is preserved, by the term of Natural Selection (Darwin, 1859, p. 61).

Variation is the feedstock for natural selection. Darwin's anatomical studies of barnacles that had occupied him for decades came to an end in 1854. His detailed monographs had earned him a Royal Medal for Natural Science from the Royal Society, but the major effect was on evolutionary theory. Darwin had observed the pervasive nature of variation and the gradual changes in structure from one generation of barnacles to another until, as Janet Browne put it (1995, p. 512), he realized that "animals became so distinct from their parents and cousins that they could be called a different species." As we shall see, evolutionary psychologists fail to take account of variation.

Darwin was attempting to explain his theory to readers who were not "natural philosophers," the cover term in that era for biologists, archaeologists, geologists, and anthropologists. The first chapter of On the Origin of Species introduces the key elements of natural selection in contexts that would be familiar to his readers. To this end, Darwin discusses the role of variation and guided selection by agriculturists who produced large juicy pears and vivid flowers, and animal breeders who bred fast racehorses and fat cattle. Darwin throughout his life assiduously sought out information from horticulturists, farmers, and animal breeders. Darwin joined pigeon fanciers' clubs and conducted his own pigeon-breeding experiment aimed at demonstrating that "long continued" selection acting on small variations in time produced breeds of domesticated pigeons so great that, "if shown to an ornithologist, and he were told that they were wild birds, would certainly, I think, be ranked by him as well-defined species" (Darwin, 1859, p. 22).

The fossil record of hominin evolution shows that Darwin was right when he stressed the role of small differences in natural selection; "slight variations" can have profound effects. The grinding action of our teeth, which acts as a food mill, increases the absorption of nutrients about five percent compared to just swallowing big chunks. A dental research group in Boston in the 1960s was testing dentures that had sharp steel edges, enabling denture wearers to easily grind up their food. But anyone who could afford to buy a second pair of false teeth would not find it difficult to buy more food, and the dentists abandoned the project. The five percent gain wasn't worth the expense and the risk of lacerated tongues. However, in the state of nature, food is not always abundant. The changes in tooth morphology that occurred millions of years ago in early hominins made it possible to absorb slightly more nutrients from a limited supply of food. Hominins thus lost the interlocking canines that prevent apes from performing the grinding motions that can mill food down into a slightly more digestible mush. Paleontologists often claim that they have found a new early hominin species based on the evidence of a single tooth.

It's easy to forget that food still is a scarce commodity in most parts of the world—my wife and I have seen horrifically stunted and wasted children in remote villages in the depths of Nepal. Life in these isolated, roadless regions is a time-capsule view of times past. If the crops fail, everyone is reduced to digging up and eating roots.

Did Natural Selection on Humans Ever Stop?

As I've already noted, according to evolutionary psychologists such as Marc Hauser, human attributes such as morality result from our having a "moral gene" that evolved and never changed sometime before 100,000 years ago, perhaps 300,000 years ago. (Hauser never explicitly gives the date.) Stephen Dawkins claims that we favor our close relations because we have a "selfish gene." And the basic premise of Noam Chomsky's widely accepted linguistic theory is that all human brains contain genetically transmitted "knowledge of language" that specifies the details of syntax of all human languages. To that end, Chomsky's views on language haven't changed since 1976, when he stated that

> language is as much an organ of the body as the eye or heart or the liver. It's strictly characteristic of the species, has a highly intricate structure, developed more or less independently of experience in very specific ways, and so on (Chomsky, 1976, p. 57).

65

In his subsequent papers and books, the specific operations that Chomsky's hypothetical language organ is supposed to carry out have changed, but it remains an innate, species-specific entity. According to Chomskian theories, the syntactic rules, speech sounds, and syllable structures of the particular language that a child hears are acquired because "parameters and principles" that activate them have been preloaded into every child's brain. In computer-speak, according to Chomsky we have preloaded software that automatically selects the appropriate instruction set for whatever language a child encounters in the first years of life. It's similar to clicking on the box for English when you open the software installation wizard, except that it's automatic—you don't have to mouse-click on "English."

In substance, Chomsky's claim is that children do not really learn language in the way that they learn to use forks or chopsticks. I have yet to see anyone proposing an "organ" that specifies genetically transmitted motor acts that enable a child to "acquire" the ability to use a fork or chopsticks, but that's precisely what Chomsky proposes for human language. If we were to substitute "table habits" for "language," a similar theory would have us believe that the maneuvers that are necessary to use chopsticks and forks derive from an innate knowledge base built into every child's brain. The "rules" for using chopsticks would be activated if a child grew up in China. The fork rules would be activated in settings in which forks were used. Different fork protocols, "fork dialects," would be preloaded into every child's brain to account for the way that forks were used in England and the United States.

There is more to Chomsky's theory—supposedly no child would "acquire" any language unless this preloaded language-software was in his or her brain. According to Chomsky and the linguists, philosophers, and psychologists who share his views, children would not be able to acquire any language because insufficient information is present in the ill-formed utterances that children hear in the early years of life. The linguistic term used to justify this claim is "the poverty of the stimulus." Since any child from anywhere can acquire any language if she or he is raised in a normal environment, then it follows that every human child must possess the same preloaded "knowledge of language" (another linguistic term). In short, no variation exists. Because there are people living all over the world, this store of "knowledge of language" must have evolved before the period in which modern humans left Africa. That must be the case because humans evolved in Africa and later colonized the rest of the world.

One recent estimate is that a number of migrations out of Africa occurred—the most recent being about 80,000 years ago (Templeton, 2002). Moreover, everyone did not leave Africa 80,000 years ago, and it is clear that the descendants of the Africans who were abducted or voluntarily migrated in the last thousand years or so have no difficulty acquiring English, Spanish, Portuguese, Turkish, or any other language. So if Chomsky is correct, the language organ, which he termed the "Universal Grammar," must have evolved and become genetically specified in every "normal" human brain well before that date. In short, since any child will acquire any language with native fluency if she or he starts before age seven years, the genetic substrate for Chomsky's hypothetical "knowledge of language" must have formed and fixed forever before the dispersal of humans from Africa, else it would be the case that some ethnic groups would be incapable of acquiring particular languages. The archaeological, genetic, and fossil evidence that will be discussed in chapters 4 and 5 pushes the date back to 250,000 years or so in the past.

Natural Selection on Humans Never Ended

But there is a problem. The only plausible mechanism, unless we really want to fall back on mechanisms related in Genesis, the Popul Vuh, or other creation myths, is natural selection. As Darwin pointed out, any biological attribute,

> if it be in any degree profitable to an individual of any species, in its infinitely complex relations to other organic beings and to external nature, will tend to the preservation of that individual, and will generally be inherited by its offspring (Darwin, 1859, p. 61).

If we read "infinitely complex relations" as the "ecosystem," which for humans is their "culture," we will be in tune with current views on the complex relations that hold between living organisms, natural selection, and the ecosystem.

We know with certainty that some aspects of human behavior have a genetic basis that resulted from localized, culturally mediated natural selection. For example, Asian ancestry often precludes adults drinking milk or eating milk products without unpleasant consequences. Asians as a group (a population, to use the biological term) tend to lack the genes that allow adults to digest milk. Not everyone can digest lactose. You can see this on the shelves of your local supermarket, which stocks lactose-free foodstuffs. Natural selection for adult lactose tolerance occurred independently in cultures in which cows, goats, or sheep were domesticated. The animals most likely were first domesticated for meat. Over time, natural selection for lactose tolerance after childhood occurred because an additional source of food was an asset in the "struggle for existence." Evolution for adult lactose tolerance, in fact, occurred on different genes in different places at different times. The most recent documented event happened in Kenya. Cattle were introduced from Ethiopia. A "selective sweep" on a gene that mutated, allowing adults to digest milk, occurred within the last 3,000 to 7,000 years. Cattle had been introduced from Ethiopia, and milk then became available for adult consumption.

The term "selective sweep" simply means that practically everyone ended up having the gene because the individuals who had this gene were more likely to survive, as well as their children. Over time, the proportion of people living in Kenya who had the adult lactose tolerance gene became overwhelming. Moreover, lactose tolerance in Kenyans and individuals whose ancestors lived in Kenya occurs from natural selection for a gene that differs from the genes that confer lactose tolerance in people who have ancestors who lived in Northern Europe (Tishkoff et al., 2007). The mutations that enabled adult lactose tolerance after childhood occurred in cultures that had domesticated herds. Natural selection for these genes increased the available food supply, thereby enhancing survival—no mystery, very simple. It is important to keep in mind that adult lactose tolerance cannot be learned or imitated—it derives from a biological trait that has a genetic basis and thus is subject to natural selection.

Humans have spread across the world to live in places that would seem improbable to their African ancestors. Tibetans have lived at altitudes over 3,500 meters and often as high as 5,300 meters above sea level for at least 5,000 years, perhaps more than 7,000 years. Surviving at these extreme altitudes means breathing air that has less oxygen. This fact has resulted in unique Tibetan biological attributes that are as unique as the Buddhist rites that mark Tibetan culture.

Tibetan culture spread during this period beyond the boundaries of present-day Tibet. My wife and I have cumulatively spent years in Nepal, India, and Tibet studying and documenting Tibetan paintings and rites. In these regions, it quickly becomes apparent when you climb over 5,000-meter-high mountain passes that Tibetans also have unique biological traits that allow them to better cope with life at these extreme altitudes. Quantitative physiologic studies show that Tibetans can function with much lower levels of oxygen in their bloodstream. The birth weights of infants born to Tibetan women in Lhasa, Tibet, at an altitude of 3,500 meters above sea level are, on average, 500 grams heavier than the babies of lowland Han Chinese women (Moore et al., 1998). Sherpas, a Tibetan group who migrated to Eastern Nepal about 350 years ago, are renowned for their ability to cope with the altitudes reached when climbing 8,000-meter-high Himalayan mountains. Over the nine years that my research group studied climbers ascending Mount Everest, it became clear that it would be impossible for virtually any "member" (the term for non-Sherpa expedition members) to reach Everest's summit without Sherpa assistance.

As chapter 2 pointed out, Mount Everest is climbed in stages by virtually everyone who isn't a Sherpa. Members first climb up from base camp at 5,300 meters to camp 2 at 6,300 meters (camp 1 is a temporary site), then at a later date from base camp to camp 2 and then on to camp 3, finally reaching camp 4 at 8,000 meters, and if all goes well ascending from camp 4 to the summit and returning to a lower camp. All of the tents, stoves, oxygen cylinders, fuel, and food are carried up and set up by Sherpas. The Sherpa ice-fall "doctors" also place ladders over the crevasses leading up over the ice fall (a moving, descending, tongue of a glacier) leading to camp 2. Sherpas carry up miles of rope and hundreds of snow stakes and ice screws, which they set up as a safety-line up the steep slope between camps 2 and 4. On a radio link one afternoon, I heard an exhausted "member" of our expedition, who had reached camp 3, fixating on hot tea. A Sherpa who also was on the radio link then climbed up 1,000 meters to camp 3 from camp 2 to deliver a thermos full of tea, and then returned to camp 2—all in a day's work.

Two independent genetic studies confirm that there is a genetic basis for Sherpas and other Tibetans being able to cope with life at extreme altitudes. Selective sweeps on the genes responsible for Tibetan extreme-altitude adaptations occurred in the last 3,000 to 4,000 years (Simonson et al., 2010; Yi et al., 2010). Their adaptation clearly is the result of natural selection acting on the pool of variation present in the human population at large. If you were to take 100 people at random off the street in Amsterdam (where the altitude is at, or slightly below, sea level depending on what street you're on), put them on a treadmill, and measure how much air they are breathing as they run, you'll find that some people need 1/10 the amount of air as other people when they run at the same speed. The differences are thought to follow from differences in the efficiency at which peoples' lungs transfer oxygen to the bloodstream (Bouhuys, 1974). In Tibetans who have been living year-round at altitudes up to about 5,400 meters for thousands of years, Darwinian natural selection has acted to favor the survival of those individuals and their children who had better respiratory efficiency.

Other recent studies point to natural selection acting on Bushmen in Africa to allow them to cope with kidney functions in extremely arid conditions and Bantu genes associated with bone density reflecting the environmental pressures of different human settings. If some attribute is essential for successful adaptation to an ecosystem, whether it's a finch or a human, natural selection will kick in. Geography, not surprisingly, plays a part in setting animals and plants on different genetic pathways. In the jargon developed by evolutionary biologists since Darwin's time, geography is a genetic "isolating mechanism." But other factors come into play in humans. Human languages and dialects act as genetic isolating mechanisms, yielding observable genetic differences in groups living in close proximity (Barbujani and Sokal, 1990; Belle and Barbujani, 2007). Put simply, men and women tend to mate and have children with someone who can understand what they are saying!

Why You Can't Speak Proper English if Your Ancestors Were Chinese

But these genetic distinctions have nothing whatsoever to do with their children's ability to learn any other language. If Chomsky were correct—that a child needs genetically transmitted (preloaded) "knowledge of language" to acquire her or his native language, we should expect to find that natural selection yielded different Universal Grammars (UGs), the preloaded knowledge bases that are necessary to acquire a language, in people whose ancestors have been speaking different languages for long periods.

Nothing in the room that you may be in while reading this book, nothing on your person, nothing that you find necessary for daily life, has an innate, genetic basis. Language clearly is the principal instrument whereby culture is transmitted. Everything that you use, wear, see, or hear is the result of cultural aggregation and cultural transmission. Virtually all of the aspects of human behavior that enhance our fitness relative to that of other species are culturally transmitted through the medium of language. Therefore, it is a certainty that natural selection would operate to create a UG that was optimized to allow a child to acquire his or her native language in the same time frame as adult lactose tolerance, high-altitude adaptation, resistance to diseases, and so on, if UG really existed. That would result in observable differences in the ability of children "acquiring" English if their ancestors spoke a language other than English for thousands of years.

Optimality Theory (OT) was proposed by linguists of Chomskian persuasion to account for how children "acquire" the phonetic structure and phonology of their native language. OT posits an innate set of principles and parameters similar to the UG, the hypothetical store of information preloaded into every human's brain necessary for a child to acquire the syntax of his or her native language. Through a process described using opaque linguistic jargon, a child activates the sounds and syllable structures that occur in the language that they hear. The store of "knowledge of language" in OT is supposedly absolutely necessary for a child to acquire her or his native language.

"Native" languages are subject to rapid change when people move. My father's native languages were Polish and Yiddish. I have command of a few Yiddish words and phrases (mostly terms describing fools and a few other phrases) and no Polish whatsoever, so Noam Chomsky could argue that my ancestors had not lived long enough in one place to have formed language organs that optimized their acquiring either Yiddish or Polish. However, that does not hold for the Chinese emigrants who arrived in California in large numbers in the nineteenth century, or any other persons of Chinese ancestry. Chinese has an attested time depth of thousands of years. The most precious holding in the National Palace Museum in Taipei is a large bronze vase bearing written inscriptions dating back to 1000 BCE. Earlier inscriptions on bones date back to 4000 BCE.

Chinese were speaking Chinese long before cattle were introduced into Kenya. If the principles and parameters of genetically specified OT were actually necessary for children to acquire Chinese, natural selection would have acted on these principles and parameters. A selective sweep or successive selective sweeps acting on the gene or genes that code OT would have yielded OT Chinese, enhancing the ability of Chinese children to "acquire" Chinese. However, the sound pattern, syllable structure, and syntax of Chinese differs markedly from that of English. Mandarin Chinese syllables, for example, all have the form consonant-vowel (CV), except for syllables that end with a nasalized consonant such as [m] or [n]. If Chinese children actually had a built in "knowledge of language" that restricted them from forming anything other than these CV syllables, they would end up being unable to say "cat" or "basket," or anything that deviated from their preloaded Chinese OT. If your parents, their parents, and so forth had lived in Iceland, where Old Norse has been preserved for 1,000 years, you would have similar problems.

When I presented this argument to a well-known MIT-trained linguist, his response was that perhaps students of Chinese ancestry at the Ivy League university where he taught did in fact have inherent difficulties "acquiring" English. I suggested that he assess the English-language proficiency of his students having Chinese ancestors. I haven't since heard anything from him.

Back to Rube Goldberg

But first, let's return to Rube Goldberg. I sometimes wonder whether some of the pundits who focus on Darwin's shortcomings ever read *On the Origin of Species*. They point out the fact that natural selection cannot account for the abrupt transitions that clearly occurred in the course of evolution. However, Darwin accounted for these abrupt transitions in 1859.

Darwin noted that natural selection can act only by taking advantage of slight successive variations; "she can never take a leap, but most advance by the shortest and slowest steps" (Darwin, 1859, p. 194). Moreover, the variations that are favored are ones that advance the survival of progeny in a particular ecosystem. The term "ecosystem" serves nicely to convey the "infinitely complex relations to other organic beings and to external nature," that to Darwin determined the selective pressures driving natural selection.

Changes in ecosystems are driven by external forces, changes in weather patterns, migrations, and so on. When an ecosystem abruptly changes, species ruled by their genes may not be able to adapt and become extinct. Most of the species that ever existed on earth are extinct; dinosaurs are only one example. But in other instances, evolution has created new species from existing ones to exploit totally new ecosystems. Everyone "knows" that fish cannot live out of water, but on the voyage of the *Beagle* around South America, Darwin observed lungfish—transitional creatures—that explained the puzzle. Deep into *On the Origin of Species*, Darwin notes that an organ might be modified for some other and quite distinct purpose. . . . The illustration of the swimbladder in fishes is a good one, because it shows us . . . that an organ originally constructed for one purpose, namely flotation, may be converted into one for a wholly different purpose, namely respiration (Darwin, 1859, p. 190).

The lungfish swimbladder-to-lung "conversion" explains why our respiratory system is as whacky as any of Rube Goldberg's machines. We have two elastic lung sacks that are inflated indirectly. Our inflated lungs increase the mass of water that we would displace if we were fish. If you were a sort of super-goldfish, this maneuver would allow you to hover almost effortlessly in a giant fishbowl. There is no direct connection between your lungs and any of the muscles that inflate the two sacks that take in air as we breathe. Instead, there is a sealed space, the pleural cavity that encloses the elastic lung sacks. Muscles in your abdomen and chest act to increase the volume of your body. This indirectly causes the lung sacks to expand, lowering air pressure in them so that atmospheric air at a higher pressure flows into them, enabling you to breathe in. The elasticity of the two lung sacks then forces air out when the inspiratory muscles relax during expiration.

You can perform the following experiment if you have a long bathtub and sharp vision. Fill your bathtub with soothing water. Then get into it, fully immersed, keeping only your head and one arm above water. Breathe in and carefully mark the water level. Breathe out, and you may be able to see the water level lower, reflecting the fishlike original function of your lungs, which was to displace water. The system makes total sense if you are a fish—increasing the volume of displaced water can enable an advanced fish to match the buoyancy of water at a specific depth, floating effortlessly. Less advanced fish such as sharks have to move constantly.

Stephen Jay Gould and Richard Lewontin reaffirmed Darwin's views in their much-cited 1979 essay, "The Spandrels of San Marco and the Panglossian Program. A Critique of the Adaptationist Programme." The provocative reference to Pangloss refers to Dr. Pangloss, the professor of "métaphysico-théologo-cosmolonigologie" who epitomizes mindless optimism in Voltaire's satiric novel *Candide*. Dr. Pangloss viewed the string of horrific events in *Candide* (based on events of the Seven Years' War and the 1755 Lisbon earthquake) as evidence that "all is for the best in the best of all possible worlds." Pangloss was mouthing the optimistic worldview of the eighteenth-century German philosopher and mathematician Gottfried Wilhelm Leibnitz [1765] (1916).

The spandrels of San Marco are an architectural feature that the master-masons and architects of Europe invented to build cathedrals that had soaring, open inner spaces. Through trial and error, features like external flying buttresses and internal arches were invented to prevent these huge stone buildings from collapsing. San Marco has internal arches whose apexes support the upper part of the cathedral's stonework. Similar arches in reinforced concrete support highway bridges today. The apex at the midpoint of the curving arch supports the roadway. In San Marco, the spaces between the arch and the sides of the building were filled in with stonework and decorated. The decorative stonework that filled in the spandrel thus was a by-product of the arch that was built for another purpose, keeping San Marco from tumbling down into a heap of large stones. The contribution to "fitness" (the selective advantage) of the spandrels wasn't to enhance the display of statuary and bas-reliefs; it was to keep San Marco from falling down. The decorative stonework thus didn't serve an "adaptionist" goal, through you could argue that art had value to the architects. Gould and Lewontin correctly pointed out that similar processes accounted for many of the features that we see in animals and plants.

Liebnitz was a setup for satire; his views on the meaning of words, which lives on in present-day "formal" linguistics, was skewered by Jonathan Swift in *Gulliver's Travels*. We'll return to that issue in chapter 4.

The roots of Jerry Fodor and Massimo Piatelli-Palmarini's book *What Darwin Got Wrong* seem to go back to their misreading Stephen J. Gould's subsequent 1985 essay, "Not Necessarily a Wing," which presented a biological example that again reaffirmed Darwinian theory. The issue that Gould addressed in his essay was the evolution of butterfly wings. Early butterflies had very small wings that served as heat-transfer devices for thermal regulation. The wings gradually became larger to enhance the efficiency of thermal regulation, but still were too small to enable them to fly. Therefore, natural selection aimed at allowing butterflies to fly by gradually increasing the size of their wings could not account for why butterflies can fly. Gould's answer was that wings gradually became larger and larger, and at some point the enlarged wings suddenly enabled butterflies to fly. Gould took care to point out that it was an example of the process that Darwin had pointed out in 1859, "an organ converted into one for a wholly different purpose." Ernst Mayr had used the term "preadaptation" to refer to this process, Gould preferred the term "exaptation."

From Motor Control to Cognition

The message of Theodosius Dobzhansky's 1973 article "Nothing in Biology Makes Sense Except in the Light Of Evolution" should now be clear. It speaks to the evolution of the cortical-basal ganglia circuits that regulate cognitive processes. Dobzhansky played a key role in twentieth-century evolutionary biology, integrating Darwinian principles with emerging genetic knowledge. Dobzhansky and his colleague Ernst Mayr (1982) pointed out the odd, seemingly inexplicable aspects of biology that made no "sense," but reflected the opportunistic nature of Darwinian evolution. It would be more "logical" and "simpler" to have a brain designed by IBM, Apple, Dell, Sony, Lenovo, and the like. As chapter 1 noted, we then would have a brain in which independent modules designed to efficiently carry out distinct tasks would each regulate some aspect of human behavior.

However, if that were so, the recurring pattern of motor, cognitive, and emotional deficits associated with aphasia, Parkinson disease, basal ganglia lesions, oxygen deficits, and other conditions would be inexplicable. Questions such as why brain damage that results in motor control problems also results in difficulty comprehending the meaning of English sentences would remain unanswered. What would motor control have to do with having difficulty understanding who kissed whom when you hear or read the sentence, "Susan was kissed by Tom"? Why do the basal ganglia, as Marsden and Obeso observed, carry out similar operations in motor and cognitive acts? The answers to these questions are clear once we realize that the neural bases of human cognition, including language, evolved by the Darwinian process by which an "organ might be modified for some other and quite distinct purpose."

The Syntax of Motor Control, Whether It's a Milling Machine or Walking

When I was an undergraduate at MIT, vacuum tubes, hot glowing globes, were the engines of all electronic devices. My classmates studying electrical engineering (MIT Course VI) and I designed and built our own hi-fi systems in our dorm rooms. The smell of solder smoldering on soldering irons was pervasive. We bought bags of World War II surplus vacuum tubes at the Radio Shack in downtown Boston. There was only one Radio Shack store then, and it was a vast jumble of electronic parts. The vacuum tubes performed very different local operations in the amplifiers and FM radios that we built. The tall 6L6s tubes were the output stages that drove our loudspeakers. The 6SN7 tubes amplified weak electrical signals that drove the 6L6s. Vacuum tubes would soon go the way of the dinosaurs, replaced by transistors that rapidly shrunk down to microscopic sizes in the integrated circuits that made the digital age possible.

While we were putting together these homemade music systems, computer-controlled machine tools were being developed in MIT's laboratories and engineering workshops. My classmates' master's and ScD theses were at the cutting edge of a technological revolution that changed the world. Computer control was more precise than the most accomplished machinists could achieve. When I enrolled in MIT's graduate linguistics program a few years later, my engineering friends pointed out that the sole difference between the syntax of the computer programs that ran milling machines and linguistic descriptions of sentence structure was that the milling machines had to work.

The demands placed on syntax do not substantially vary whether the task involves running a milling machine, walking, sentence structure, or any form of serial behavior. Karl Lashley, one of the pioneers of neuroscience, pointed this out:

> Temporal integration is not found exclusively in language; the coordination of leg movements in insects, the song of birds, the control of trotting and pacing in a gaited horse, the rat running the maze, the architect designing a house, and the carpenter sawing a board present a problem of sequences of action . . . each of which similarly requires syntactic organization (Lashley, 1951, p. 113).

Walking, for example, involves positioning your foot to achieve heel strike at the precise moment of contact, whether you are on smooth pavement or the irregular surface of a country path. Your heel must flex at the proper moment, and that moment is different depending on whether you are on an incline, your posture, and your speed. Many linguists mistakenly think that the rules of syntax that, for example, govern word placement in an English sentence differ from motor control because their linguistic rules involve "selectional constraints." The linguistic jargon term "selectional constraints" simply means that the rules have to be applied in a certain order. That clearly is also the case for the sequence of "rules" that governs walking. You cannot flex your foot to achieve heel strike before your foot is about to hit the ground. And if you are running, a different set of motor control rules and selectional constraints comes into play. As Marsden and Obeso (1994) presciently observed, the "rules" that the basal ganglia execute during motor control are not different from those necessary to describe cognitive processes. An organ that we share with frogs has been adapted to serve new ends.

Other Reshaped "Organs": Three Little Bones

"A very large snake is eating my boyfriend on the arm." It probably was the strangest call that the 911 operator in Putnam, Connecticut, had ever heard. Karen Ziner in the *Providence Journal-Bulletin* (October

14, 1998) reported the bizarre event. Christopher Paquin's 19-foot, 260-pound pet python had twisted her coils around him, pinning him to the floor of his knotty-pine paneled sunroom. The snake opened its hinged jaw wide and swallowed his arm. Paquin called for help; his girlfriend, Tammy Breton, tried to pull the snake off using visegrip pliers, but Paquin yelled that she was hurting the snake. And so she dialed 911. When the police arrived, they found Paquin "being devoured all the way close to the shoulder." Three policemen managed to separate Paquin from his pet. The snake was duly dispatched.

Paquin's pet snake was able to engulf his arm because snakes have jaws that are connected by three hinge bones that enable them to dramatically open their jaws wide. There are other instances of pet pythons eating their owners and small children. You can find three similar bones in your middle ear. Resized and reshaped, they serve as a mechanical sound-amplifying system in all mammals. Mammals also possess the paleocortex (old cortex) that also differentiates mammals from reptiles. The paleocortex includes the anterior cingulate cortex (ACC). As chapter 1 noted, the anterior cingulate cortex plays a role in directing attention to virtually anything that you wish to do. The ACC dates back to Therapsids, mammal-like reptiles who lived in the age when dinosaurs roamed the earth.

The soft tissue of the brain obviously hasn't survived 260 million years, and the inference that Therapsids had an ACC is based on their fossil remains having the three middle ear bones found in all present-day mammals. The initial function of the hinge bones of the reptilian jaw was to open the jaw wide. In the course of evolution, the hinge bones took on a dual role, also functioning as "organs" of hearing. The final mammalian transition involved the former jaw bones migrating into the middle ear, where they serve as a mechanical amplifier, increasing auditory acuity. In the Darwinian "struggle for existence," being better able to hear her babies enhances the possibility of a mother keeping in contact with her suckling infants. This fits in with mammals having a paleocortex. Lesion studies show that the ACC's role in a mouse mother is to get her to pay attention to her pups. Mouse mothers don't pay attention to their infants when neural circuits involving the ACC are disrupted (Maclean and

Newman, 1982; Newman, 1985). Similar, but more general, problems in maintaining attention occur when ACC to basal neural circuits are degraded in humans. Patients become apathetic, they don't attend to events in the flow of life when the ACC to basal ganglia circuit is degraded (Cummings, 1993).

Evolutionary tinkering in the best tradition of Rube Goldberg converted the ACC's role in maintaining a Therapsid or mouse mother's attention to her infants to directing attention to anything. Virtually every PET or fMRI neuroimaging study ever published shows ACC activity when the subjects are asked to perform any task. Another reminder of Rube Goldberg engineering and our reptilian heritage is that you may develop an earache because you grind your jaws as you sleep. We retain the "old" nerve pathways between the jaw and the bones that have migrated into the middle ear.

The ACC has another role in mice and presumably Therapsids that enhances mother-infant interaction. The ACC controls the laryngeal "mammalian isolation cry"—the cry that keeps parents awake for months after a baby is born. It is part of a neural circuit involving the basal ganglia. In human adults, that circuit continues to be involved in laryngeal control. "Hypophonation," exceedingly quiet speech from laryngeal dysfunction, is a sign of Parkinson disease, which degrades basal ganglia activity (Jellinger, 1990). As noted earlier, Parkinson disease also can result in speech production errors involving coordinating laryngeal, lung, tongue, and lip activity (Lieberman et al., 1992). The large bilateral basal ganglia lesions of patient CM, who was studied by Emily Pickett and her colleagues, resulted in highly aberrant laryngeal control. Phonation erratically occurred in consonants that should not have been phonated, indicating serious problems in laryngeal muscle control (Pickett et al., 1998). The afflicted members of the KE family whose basal ganglia development was affected by a broken FOXP2 gene also have speech production errors.

The inference drawn between anatomy and brains—the former hinge bones of reptiles and the anterior cingulate cortex—thus allows us to trace the evolution of one of the neural capacities that allow us to talk back to Therapsids.

Talking and Breathing

The opportunistic fish-based evolutionary process that accounts for our lungs also affects the way that we talk and sing (or for coyotes and wolves, howl). It necessitates complex planning when we talk or sing because the alveolar (lung) air pressure would start out very high and then fall as the two elastic lung sacks deflate. This is the case during quiet respiration; the energy that pushes air out of your lungs during expiration has been stored in the elastic lung sacks. The elastic lung sacks act as springs, exerting maximum pressure when distended that gradually falls. It's as though you blew up a balloon and then let go. The balloon would start by flying around fast, then slow down, and finally fall to the ground as it deflated.

If we didn't act to modulate the alveolar air pressure pattern that occurs during quiet respiration, talking and singing would be impossible. That's because the rate at which the vocal cords of the larynx move, which determines the fundamental frequency of phonation (F_0), the perceived pitch of your voice, depends on the alveolar air pressure. When you intend to speak a long sentence or sing a long musical phrase, you have to take lots of air into your lungs before you utter a word, distending the elastic lung sacks. If you did nothing, the alveolar air pressure would be so high at the start of the sentence that your vocal cords would just be blown apart, producing a raspy sound. As you continued to talk, the alveolar air pressure would fall and the pitch of your voice would fall precipitously. What you must do instead before you utter a word is "program" a muscle command function that starts by maximally opposing (holding back) the elastic lung sack spring force and that gradually falls to match the falling lung sack elastic force, to achieve a level alveolar air pressure. The elastic lung sack hold-back function depends on the length of the sentence that you intend to say.

Weird and complicated? It's the result of evolutionary tinkering with swim bladders to "make" lungs.

Chapter Four

Chimpanzee Brain 2.0

My wife Marcia's mother, Mina Mazo, landed on Ellis Island in 1913. Her ancestors had lived in Spain for centuries before they migrated to Russia, but her face seemed to have a bit of Mongol. The Golden Horde invaded in 1251 and controlled large parts of Russia for centuries. So whenever Marcia reads about some celebrity's DNA showing that she or he had a surprising ancestor, she wonders whether she has a trace of Genghis Khan!

It is now commonplace to read of DNA genotyping being used to establish paternity, to identify criminals, or to absolve people falsely accused of crimes. DNA genotyping in 1998 confirmed that Thomas Jefferson fathered six children with his slave Sally Hemings. Perhaps more surprising was the comparison of the human and chimpanzee genomes in a report published in the journal *Nature* in 2005. Our genes are almost identical. Depending on how differences are calculated, only one or two percent of our genes are different. But another, more surprising, result of current genetic research may account for the fact that although chimpanzees, our closest living relatives, are far more capable than most people believe, we are not chimpanzees. As we'll see, we have a supercharged chimpanzee brain—one that can access more information, is better at learning things, and most importantly, has an exceptional creative capacity. But superchargers don't work in isolation, whether the engine is a car's or a brain's. Our brains also can store more information.

Big Brains

The first anatomical study of an ape, Tyson's 1699 dissection of an orangutan, revealed many of the anatomical affinities between great apes and humans, but it clearly showed that the ape brain was much smaller than that of any normal human adult. Much attention has since been focused on our having brains that are about three times bigger than a chimpanzee brain because brains require lots of biological support to operate. Hence, having a very large brain surely signifies that it's useful. A staple of books on human evolution is the chart showing an increase in homimin brain size over time.

Current quantitative assessments show that we have a scaled-up primate brain that has about three times as many neurons, the basic computing elements of all brains, as a chimpanzee brain (Herculano-Houzel, 2009). Some hints as to what this signifies can be drawn from the fact that the size of most parts of the mammalian brain scale up in proportion to overall brain size (Finlay and Darlington, 1995), but the human brain differs in that the posterior temporal cortex is disproportionately larger than would be expected (Semendeferi et al., 1997, 2002). Temporal cortex, located above your ears, is part of our long-term information storage system. Working memory, discussed earlier, appears to access information from permanent memory through neural circuits linking prefrontal cortex to temporal cortex and other structures, in much the same manner as information is accessed from the hard drive of a digital computer (Badre and Wagner, 2006; Postle, 2006).

Our disproportionally larger temporal cortex thus signifies natural selection for enhanced memory, but the situation is a bit murky because the human hippocampus, which is involved in consolidating memories ("saving" information, to use another computer analogy) is disproportionately smaller. However, the smaller relative size of the human hippocampus might reflect its evolutionary history, which involves navigation and relatively lower demands on its role in finding your way around. When you take a walk, or drive to some destination, you may not be consciously aware that you are storing landmarks that will enable you to retrace your path. The hippocampus has that role in animals ranging from mice to man. The hippocampus codes the "implicit" memory traces that enable people and mice to retrace their steps. The hippocampus is involved in making unconscious mental maps—finding your way around.

It's striking to note that exercising your hippocampus seems to make it larger. Conversely, having other ways of finding our way around might account for it being proportionately smaller in humans. The street plan in many parts of London resembles a maze. On a trip to London, my wife and I spent almost an hour looking for the pub in which we had rented a room, though it should have been a 15-minute walk from the tube station. The room was great and the pub served excellent meals, but finding it was a chore. Many of London's medieval wood structures were destroyed by the Great Fire of 1666, and new streets were laid out in an orderly grid pattern. But streets still twist and turn, and small squares break up the grid pattern. Maguire et al. (2006) found that London taxi drivers, who have to find their way through this maze, had a larger hippocampus than most other people. More experienced drivers had a larger hippocampus than novices. In contrast, time on the job had no effect on the size of the hippocampus for London bus drivers, who negotiated fixed routes that didn't tax their brains. One of my former students, Jennifer Adeylott, who now teaches at Birkbeck College in London, wants to replicate the Maguire et al. study. In the last decade, London's taxi drivers have been using GPS navigation systems. Jennifer predicts that in time, London taxi drivers won't have a larger hippocampus than bus drivers. Our relatively smaller hippocampus compared to its ape-brain proportion may simply reflect it being large enough for us to find our way about in most situations even when we aren't using a GPS navigation system, its original function, and also saving other information. As Dobzhansky noted, "Nothing in biology makes sense except in the light of evolution."

The human prefrontal cortex has long been associated with "higher" cognition, and as the studies mentioned in the previous chapter (and hundreds more) show, linked with the basal ganglia to other neural structures, it is involved in the range of cognitive acts involving "executive control." Prefrontal cortical areas, dorsolateral, and ventrolateral, work through the basal ganglia as well as through direct neural circuits with information-storing regions of the brain, pulling memory traces of images, words, and probably other stored information, into short-term, working memory (for example, Postle, 2006; Badre and Wagner, 2006; Miller and Wallis, 2009). Terrence Deacon in his 1997 book *The Symbolic Species* claimed that humans had a disproportionately larger prefrontal cortex that accounted for our enhanced cognitive ability. Deacon's claim was based on comparisons of MRIs of human brains and a small sample of ape brains. However, as Semendeferi et al. (2002) pointed out, the MRIs inherently cannot show that humans have a disproportionately larger prefrontal cortex. The human frontal cortex, which includes the prefrontal areas of the cortex as well as areas involved in motor control, is not disproportionately larger than an ape's, and it is impossible to differentiate prefrontal cortical areas from the motor regions of the frontal cortex on an MRI. The situation is murky, as is the case for many questions concerning how brains work.

Factoring Out Body Size

But even if the human prefrontal cortex is not disproportionately large, it still is three times as big as a chimpanzee's. Stephan and his colleagues used anatomical dissections to compare the size of various parts of the brain in primates and concluded that prefrontal cortex was about three times larger in humans than chimpanzees (Stephan et al., 1981)—the general size increase of the human brain. In assessing whether this or any measure of brain size is significant when making inferences about intelligence, body size has to be factored out. A bigger brain doesn't necessarily indicate greater intelligence, because animals that have bigger bodies have bigger brains. Whales, for example, can have brains that are five times larger than a human's. Whales are intelligent animals, but they are not five times more so than humans. Jerison (1973) addressed this problem when he introduced the encephalization quotient (EQ), which attempted to factor out body mass to provide an estimate of brain mass devoted to cognition. Computing the EQ involves many assumptions about the relations that hold between body mass and the brain.

Having stated that brain size, in itself, isn't a measure of cognitive ability, it is clear that the probable information storage capacity of hominin brains has dramatically increased since we split off from apes. Both various EQ measures and absolute brain size increased over the course of hominin evolution. There isn't evidence for a very large increase in hominin brain size until fairly recently. By "recent," I mean one million or so years ago. Lucy, *Australopithecus afarensis*, described by Don Johanson in his 1981 book, lived about three million years ago. Lucy's brain is only slightly larger than a chimpanzee's, whether brain mass or EQ estimates are considered. Lucy could walk upright, but there are no signs that she or her close relations achieved much more than chimpanzees. The first evidence for hominins having achieved anything more complex than chimpanzees is associated with *Homo habilis*, often identified as the first hominin species that made stone tools. *Homo habilis* had brains that were about 60 percent bigger than a chimpanzee's. It isn't until one million years ago that brain size in *Homo erectus* began to increase substantially. Daniel Lieberman's book *The Evolution of the Human Head* discusses the evolution of brain size in detail (D. Lieberman, 2011, pp. 192–204 and 544–549).

What Drove Increases in Brain Size?

Bigger brains require more energy to run them, so it is reasonable to infer that the evolution of bigger brains must have conferred abilities that allowed our distant ancestors to prosper in, as Darwin put it, "the infinitely complex relations to other organic beings and to external nature," that all beings face (Darwin, 1859, p. 61). But there isn't anything special about hominin brain size increasing over time.

Lartet in 1868 documented a brain-size race between wolves and their meals. Lartet examined the skulls of wolves and sheep buried in the layers of sedimentary earth deposited by the river Seine. The depth of each layer provided a measure of time. Animal remains that were found in a deeper level lived at an earlier time. Wolves are cleverer than sheep. Not surprisingly, the wolves living at any given time had bigger brains than their meals. However, Lartet discovered that the brains of sheep closer in time to 1868 were bigger than the brains of wolves that had lived at earlier periods. The brains of both wolves and sheep had increased in size over time. The wolf-sheep brain-size race makes sense if having a bigger brain increased cognitive ability in both sheep and wolves. Cleverer sheep had sometimes outwitted wolves and survived. As Darwin had predicted, the offspring of these sheep retained their father's and/or mother's bigger brains. But as the brain size of the menu increased, a premium on bigger brain size occurred in the diners, so wolves maintained their edge.

Food, and not being food, was the factor that drove the wolf-sheep brain-size competition. We can account for the evolution of larger brains in wolves simply on the basis of success in obtaining more food enhancing the survival of an individual wolf and his or her progeny in the Darwinian "struggle for existence." Other factors must be involved in the pace and extent of the evolution of hominin brain size.

Many attempts have been made to answer this question. One recurring theory hinges on abrupt climate changes taking place in Africa, where early hominins evolved. William Calvin (2002) suggested alternating periods of frigid glacial cold and heat that forced our distant ancestors to think of creative solutions to survival. However, there is no evidence for glacial cold in Africa, which poses a problem for Calvin's theory. Another proposal in the same vein suggests alternating periods of drought and heavy rainfall that resulted in alternating desert-like or lush rain forests in the Rift Valley of Africa as the causal element driving hominin brain-size enlargement. However, archaic hominins migrated all over Africa and most likely would have moved away when the climate became inhospitable. Modern humans in like circumstances will move if they can do so. The pictures taken by the photographers of the Farm Security Agency in the 1930s of Dust Bowl families on the move in their trucks to California bear witness to that recent mass migration.

Other species have acted in the same manner, though they didn't have trucks. The migration of Mammuts, tusked cold-weather-adapted elephant-like animals, fleeing from the advancing glaciers in the last ice age from northern Europe, can be traced by their fossil remains. And if weather, in itself, caused hominin brain size to increase, why didn't brain size increase to the same degree in other species? Some genes have been identified that are associated with aberrantly smaller human brains, but the genetic mechanisms responsible for the enlargement of the human brain remain unclear.

The Brain's Dictionary

Attics are where you put the odds and ends that collect as life goes on and that might be useful someday or other. The posterior parts of your cortex, the parietal and temporal lobes, are a sort of mental attic, connected to subcortical structures such as the hippocampus and cerebellum, as well as to prefrontal cortex. It is an attic in that nothing seems to be stored in neat rows of boxes, books, or folders arranged on some neural equivalent of library shelves. Neuroimaging studies exploring the manner in which we attain and store knowledge show that it seems to be a matter of perceiving and storing a jumble of bits and pieces. One of the surprising findings of fMRI and PET studies that map brain activity when subjects look at pictures or read words is that when we see a picture of an elephant or read the word "elephant," cortical areas that code colors and shapes are activated. The activated areas do not resemble a complete photographic-like image of an elephant, they instead code bits and pieces—colors and elementary shapes, lines and angles, that somehow (no one knows how) are assembled in the mind's eye as an image of an elephant.

Independent studies have identified areas of the cortex that are activated in both perception and imagery (for example, Martin and Chao,

2001; Kosslyn et al., 2001; O'Toole et al., 2005). The pattern of activation also extends to cortical areas involved in perceiving motion when a movement is involved (Kourtzi et al., 2002), and even to neural structures governing motor control when the word's referent involves overt motor activity, for example, "hammer" (Martin et al., 1995). Various areas of the human visual system are specialized for the perception of different visual attributes, color, shape, as well as motion. These memory traces are clearly formed by life's experience, bits and pieces in different heaps, put together somehow in a neural attic. Subcortical structures such as the hippocampus, which seems to solidify the neural representation of words, as well as the knowledge-base that allows you to find your way hack home, are part of this network. Laboratories, such as Randy Buckner's at Harvard University, are focusing on the possibility that degeneration of neural pathways to the hippocampus plays a role in Alzheimer's dementia (cf. Buckner, 2010).

In short, the expansion of the posterior regions of the human brain can be viewed as enhancing our brain's dictionary—its store of knowledge. A vastly larger knowledge base can enable its owner to consider more alternatives in solving a problem, considering a different course of action, or in implementing a creative impulse. However, in itself, a large store of information cannot account for why we are the most creative, hence unpredictable, living species. Inheriting a large library doesn't in itself lead to writing the great American novel!

Learning Words

It also seems clear that the particular activation patterns observed in neuroimaging experiments are determined by a person's experience. One of the subjects in the studies reviewed by Martin and Chao (2001) was an Indian mahout who routinely worked directly with elephants, guiding them with a prod, sitting atop them, and so on. Hence, the activation patterns for the word "elephant" and a picture of an elephant included color and shape, but no motor control sites. I suspect that an Indian mahout, if placed in the fMRI magnet, would show activation in neural structures involved with motor control. I also wonder whether the meaning and activation pattern would be different, or less robust, for words whose meaning was derived by reading, seeing a movie, or hearing someone telling a story? The semantic referents, meanings, conveyed by a word are determined by an individual's knowledge-base, which for practically everyone is more extensive than that which follows from their direct sensory experiences. What would you make of the phrase, "slam the dog home?" Is it man's best friend that is being slammed? In Joseph Conrad's novel *Lord Jim*, the "dog" in question was a large metal cam that could be pounded into place with a hammer to seal a ship's watertight door. Would neural manual motor-control activation similar to that for the word "hammer" occur if I were a subject in an fMRI experiment and read or heard the word "dog"?

As we proceed along the path of life, words stored in our brains evoke memories of events, embedded into a richer context. When I sometimes hear or read the word "bicycle," what comes into my mind's eye is my father, running behind me, keeping my bicycle upright on my first attempt to ride. Is it a definition of the word "bicycle"? Not really, but it's part of what "bicycle" sometimes brings to mind to me when I think about bicycles. The meaning of a word is always linked to a time and a place. The mathematician and philosopher Jacob Bronowski (1978) pointed out that "atom" had a different meaning to a physicist in 1960 than to another physicist living in 1890, and quite another meaning to the Greek Democritus in the fourth century BCE, to whom atoms were the indivisible elements of matter. Words take on different meanings and clearly can mean something very different to different people in a given time and place. Bronowski stressed the fuzzy elastic nature of words that eludes "precise" definitions. He noted that you can carefully define chairs and tables in terms of their physical properties, but at the end, someone can declare that s/he is using this chair as a table.

The indeterminate fuzzy referents of words constitute an element of their utility. They adapt and take on new meanings as it becomes necessary or useful to code a new concept into the shell of a word. A glance at the entries in the unabridged edition of the *Oxford English Dictionary* will reveal sentence after sentence in which a word's "meaning" changes. That also is the case for chimpanzees accultur-ated to communicate using sign language. Washoe, one of the chim-panzees raised by Alan and Trixie Gardner from infancy using Amer-ican Sign Language (ASL), learned to produce about 150 words. In some instances, she formed compound words to describe new events. "Water-Bird" was how she described a duck. In one documented in-stance, she extended the semantic referent of "dirty." Washoe, like the other young Gardner chimpanzees, at first wore diapers. The word "dirty" was used to refer to her soiling herself. But when Roger Fouts, the graduate student working with Washoe refused to give her a sec-ond slice of cake at a party, Washoe signed "dirty Roger." I suppose that an evolutionary psychologist could use that episode as evidence for the innate nature of epithets! Washoe passed away on October 30, 2007, at age 42 years, signing words to the end, both to herself as she "read" picture books and as she communicated with humans and her sign-language-using chimpanzee companions.

Other views on what words signify have been the stuff of philos-ophers and satirists. Wilhelm Gottfried Leibnitz was a perfect tar-get for Jonathan Swift. Leibnitz [1765] (1916) commented on the imperfection of words, their lack of precise meaning, in a critique aimed at John Locke. Locke had proposed that we are born with-out innate views and form concepts through experience, and hence words would have different meaning to different people. Locke's po-sition wasn't acceptable to Leibnitz, whose critique was completed in 1710, but wasn't published until 1765. However, the debate must have been in the air. In Jonathan Swift's *Gulliver's Travels*, the savants of the Academy of Lagado have

> a scheme for entirely abolishing all words whatsoever. . . . An ex-pedient was therefore offered, "that since words are only names for things, it would be more convenient for all men to carry about them such things as were necessary to express a particular business they are to discourse on." . . . Many of the most learned and wise ad-here to the new scheme of expressing themselves by things; which has only this inconvenience attending it, that if a man's business be very great, and of various kinds, he must be obliged, in proportion,

to carry a greater bundle of things upon his back, unless he can afford one or two strong servants to attend him. I have often beheld two of those sages almost sinking under the weight of their packs, like pedlars among us, who, when they met in the street, would lay down their loads, open their sacks, and hold conversation for an hour together; then put up their implements, help each other to resume their burdens, and take their leave (Swift [1726] (1970), p. 176).

Are Words and Concepts Innate?

Leibnitz argued for a store of innate concepts in the human mind that would determine what words conveyed. That argument is still with us. Noam Chomsky (2000, pp. 65–66) has expressed that view. Jerry Fodor (1998) enlarged on the theme. Fodor thinks that the meanings of about 50,000 words are innately determined, preloaded in our brains. Few people could hold to such a zany idea. Is "bulldozer" an innate concept? Was "bulldozer" an innate property of the human brain in the year 1850?

Steven Pinker (2007) instead modestly proposes that the human brain has an unspecified, smaller number of "universal elements of thought." The French epigram attributed to John-Baptiste Alphonse Karr in 1849 sums it up "Plus ça change, plus c'est le même chose"—nothing changes. Pinker's short, incomplete list of "basic concepts" includes state, thing, path, property, manner, acting, going, being, having, on, at that "shape our understanding of the physical and social worlds" (pp. 80–82).

Getting off the Language Hobbyhorse

One more swipe at poor Leibnitz is in order. It came to me, or rather was thrust in my nose one night as I was reading one of Patrick O'Brian's meticulously crafted historical novels. The O'Brian novels track the adventures and misadventures of Dr. Stephen Maturin and Captain Jack Aubrey of the British Royal Navy during the Napoleonic wars. Maturin is an Irish-Catalonian medical doctor who is also a master spy in the British service. O'Brian interlaced real historical figures such as Sir Joseph Banks, a leading figure in British science, and the details of naval engagements from official records into the adventures of his fictional characters. At one point, Maturin, who is the author's voice, is attending the opera in London between naval adventures. He is suddenly conscious of Diana Villiers, whom he desperately loves, sitting apart from him in a group below his private box. He senses the perfume that he had given to her not long before she moved away from him. O'Brian gives Maturin these thoughts,

> A foolish German had said that man thought in words. It was totally false; a pernicious doctrine; the thought flashed into being in a hundred simultaneous forms, with a thousand associations, and the speaking mind selected one, forming it into the inadequate symbols of words, inadequate because common to disparate situations— admitted to be inadequate for vast regions of expression, since for them there were the parallel languages of music and thought. . . . He himself at this moment was thinking of scent (O'Brian, 1972, p. 470).

It is clear that we can think outside the box of language. I started to take photographs when I was 13 years old. Through odd jobs and photo-contest winnings, I bought my first Leica three years later. When I take a picture, it's a mode of expression, thought without words. I avoid conscious thought because I then immediately begin thinking in words, disrupting the link between me and what's out beyond the viewfinder. Words are only one of the forms by which we express our thoughts. Linguists often seem to think that language is the only mode by which we think.

What Might Be Different about the Human Cortex?

There's no disagreement in the neuroscience community about whether the human prefrontal cortex plays a significant role in differentiating people from other animals. Absent prefrontal cortex, basal ganglia operations would be limited to motor control and some aspects of emotion. But there has been a lopsided focus on the cortex in attempts to account for the brain bases of being human. I have departed from that paradigm, discussing cortical-basal ganglia circuits and the local operations of the basal ganglia. But prefrontal cortex cannot be ignored, and this raises the question of whether any unique attributes of the human prefrontal cortex account for human creativity.

One possibility is that there is something very distinctive about the structure of human prefrontal cortex. As chapter 1 pointed out, different areas of the cortex are often identified using the "cytoarchitectonic" maps produced by Brodmann at the turn of the twentieth century. The cortex (most accounts use that term to refer to the neocortex) is made up of six layers of neurons. Brodmann found that the distributions of various types of neurons in these layers differed in adjacent areas of the cortex. Neuroscientists have since attempted to discover the functional difference of these differences. Thus another explanation for why we can talk and behave in very unusual ways compared to other primates might rest on human cortical areas having unique structural characteristics. However, this does not seem to be the case. Petrides (2005) in his review of both classical and current studies, including his extensive contributions, concludes that "the basic architectonic organization" (the distribution of neurons in the frontal layers of the cortex) is the same in humans and monkeys.

Petrides's assessment of the then available data doesn't differ from that of other recent studies. Current studies indicate that the posterior upper (dorsal) cortical areas are involved in motor control. Ventrolateral prefrontal cortex is involved in actively selecting and retrieving information stored in other regions of the brain accordingly, connected to posterior regions by circuits that involve the subcortical basal ganglia. Dorsolateral prefrontal cortex is involved in monitoring motor or cognitive events during a task, taking into account previous experience. Petrides's assessments fit in very well with the local operations of the basal ganglia noted by Marsden and Obeso (1994) and the findings of neuroimaging studies such as those of Oury Monchi and his colleagues, discussed in chapter 2.

The Monchi group showed that ventrolateral prefrontal cortex and the caudate nucleus were active when effecting a change in direction of a thought process, comparing different cognitive criteria and when shifting to a new criterion. The caudate nucleus and putamen of the basal ganglia were active in different phases of this process. The basal ganglia essentially act as a sequencing engine that can shift to a different sequence of motor or cognitive acts. It's pertinent to again point out that creativity depends not only on cognitive flexibility, but also on having an active cortical memory that allows a person to store different concepts and plans of action. That entails having a human cortex that can store a range of alternative concepts and actions—a big cortex. The threefold increase in the size of the human brain and the relative enlargement of temporal cortex thus is significant. The basal ganglia appear to be the "switching" mechanism, but cortex is the store in which we place and access alternative concepts. Katerina Semendeferi and others also have suggested that the efficiency of information transfer in cortex in humans has been enhanced. That seems probable in light of the ongoing genetic studies that we will discuss later.

Do We Have Unique Neural Circuits?

This brings us to a mystery. Scores of independent studies point to neural circuits linking cortex and the basal ganglia being the engines that regulate internally guided motor acts, being involved in learning complex motor acts, and conferring "executive control"—cognitive capacities such as working memory and cognitive flexibility. The root of creativity is cognitive flexibility—selecting a different way of thinking about something or doing something different from what was expected. What's mysterious is that our cortical to basal ganglia circuits don't seem to differ from those of monkeys and apes. Tracer studies of monkeys and other species and diffusion tensor imaging (DTI) of human brains do not show any marked differences. Neurotransmitters seem to be the same. However, apes cannot learn to talk, and though they can operate keyboards, it is not a chimpanzee who is writing this book. Nor could a chimpanzee read this sentence. There must be some neural distinction between human and nonhuman primate brains. Something has to be different.

One reasonable possibility is that humans have unique neural circuits that allow us to talk, acquire language, and think. Terrence Deacon in his 1997 book *The Symbolic Species* proposed that humans have a unique, species-specific, neural circuit that directly links the cortex to the larynx. As noted earlier, the larynx is not in itself the key to talking. Talking involves learning and being able to execute complex coordinated maneuvers involving the tongue, lips, soft palate (which seals off the nose), larynx, and lungs. However, Deacon seems to believe that this species-specific neural circuit plays a central role in speech, language, and cognition. W. Tecumseh Fitch in his 2010 book *The Evolution of Language* has gone further. Fitch explicitly states that this neural circuit is the key to vocal imitation, which he implies is the means by which children learn to talk. Fitch also endorses Noam Chomsky's views concerning Universal Grammar, so it isn't entirely clear whether Fitch thinks that vocal imitation in itself would suffice to "acquire" language. However, the hypothetical unique neural circuit somehow seems to be linked to the most recent version of Noam Chomsky's language organ, the "Faculty of Language, Narrow" (FLn). The FLn is one of the current candidates of Chomsky's followers for our ability to form and comprehend complex sentences. We'll return to Chomsky's views on language in chapter 6, but the question here is, does this unique human neural circuit really exist?

The basis for the Deacon-Fitch claim goes back to one of the first attempts to map out human neural circuits. Kuypers in 1958 attempted to adapt a lethal tracer technique used in animal studies to study human neural circuits. The Nauta-Gygax procedure had been used to trace out neural circuits in animals. The first step entails destroying a discrete part of an animal's brain. The animal then is allowed to live for a few weeks. The next step entails "sacrificing"— aka killing—the animal. The animal's brain then is impregnated with a silver solution that highlights the structure of neurons. The brain then is sliced up and examined microscopically. If everything works out, the silver solution shows damages to "downstream" neurons that were connected to the part of the brain that had been destroyed, thereby tracing out a neural circuit.

Kuypers attempted to ethically use this technique to map out human neural circuits in patients who had died after strokes that had damaged one hemisphere of their brain. He claimed that he had found a circuit that directly linked part of the cortex to the neurons that control the muscles of the larynx. In 1990, Iwatsubo and his colleagues used the Nauta-Gygax technique to study the brain of an 84-year-old woman who died after suffering two massive strokes. They reported degeneration in spinal cord neurons that they believed showed a direct human cortical to brainstem neural circuit.

Uwe Jurgens, whose 2002 review article attempted to pull together the results of 301 studies on the neural bases of motor control, is also often cited to support the claim that humans have this hypothetical unique neural circuit. Jurgens presented no evidence other than citing the Iwatsubo study in a brief comment. Jurgens apparently overlooked the section of the Iwatsubo paper that documented the extensive brain damage that the patient had suffered after two massive strokes. Moreover, Jurgens was unaware that the conclusions reached by Iwatsubo had been refuted by a subsequent study. Terao et al. (1997) attempted to replicate Iwatsubo's findings. The brainstem neurons of four patients who died after brain damage were compared to those of four age-matched patients who had died from diseases that don't involve the brain. No difference was apparent when the brainstem neurons of the two groups of patients were examined. The neuronal degeneration observed by Kuypers and Iwatsubo and his colleagues thus didn't signify downstream damage to brainstem neurons linked to the cortex. The changes in brainstem neurons observed by Kuypers and Iwatsubo instead may reflect general neural decay after death.

Even if the brainstem damage seen by Kuypers and Iwatsubo and his colleagues had reflected downstream damage to neural structures, the finding could not have shown that a direct cortical to larynx circuit existed. No one dies because the vocal cords of their larynx cannot phonate. In the case of the Iwatsubo study, CT scans and an autopsy showed that after suffering two strokes, "massive infarctions of the entire territories of the middle cerebral artery on the right and anterior and middle cerebral arteries on the left had occurred." The midbrain and brainstem neural structures that control respiration, blood circulation, and other life-supporting systems were destroyed, as well as the basal ganglia and pathways to the basal ganglia.

Jurgens, perhaps overwhelmed by the flood of data of the 301 research papers that he reviewed, also overlooked the findings of the Pickett et al. (1998) study that he cited, but didn't take into account. As noted in chapter 2, Emily Pickett and her colleagues studied the speech, cognitive, and sentence comprehension deficits of a woman who after suffering a coma, ostensibly spoke Irish. The Pickett study showed that damage limited to the basal ganglia profoundly disrupted laryngeal speech motor control, as well as cognitive deficits. Phonation was irregular, suddenly increasing and decreasing. Laryngeal phonation often occurred where it wasn't supposed to be produced. The patient had difficulty coordinating laryngeal inactivity with lip and tongue movements to produce consonants like [p], [b], [d], [t], [g], and [k].

Similar problems occur in Parkinson disease, which always causes motor deficits (Lieberman et al., 1992). "Hypophonation"—pathologically quiet laryngeal phonation—also occurs in Parkinson disease (Jellinger, 1990). If we humans actually had species-specific neural circuits that bypassed the basal ganglia and directly linked motor cortex to the larynx, then neither the patient studied by Pickett nor Parkinson disease patients would have any speech motor control problems. Nor would speech and cognitive deficits occur in patients where brain damage was limited to the basal ganglia and circuits to it in cases of aphasia—permanent loss of language and speech as Naeser et al. (1982) showed. In short, humans do not have a unique cortical to laryngeal neural circuit.

Subsequent studies have used noninvasive techniques to study the neural circuits that control vocalization in living humans. Schulz et al. (2005), for example, used positron emission tomography (PET) to discern the neural circuits that are active when people vocalize. They concluded that we use the "phylogenetically older system present in all mammals—neocortical and subcortical motor regions." DTI, which directly maps out neural circuits, confirms that we have the same circuits present in monkeys and apes (Lehericy et al., 2004). At the end of

Jurgens's review article (Jurgens, 2001, p. 251), like Cummings, Alexander, Graybiel, and other neuroscientists, he concludes that "motor coordination of learned vocal pattern . . . comes from the motor cortex and basal ganglia." Fitch, who in his talks, papers, and 2010 book leans on Jurgens's review article, may not have read to the end of the article.

In short, we don't have a unique neural circuit that is the key to human speech and language. So we're back to the central mystery. We and other primates have similar neural circuits.

Cortical-Basal Ganglia Circuits — The Engine of Human Unpredictability and Creativity

The range of independent studies that we have reviewed points to neural circuits linking activity in prefrontal cortex with the basal ganglia and other parts of the brain, yielding human cognitive flexibility. There are, as I have pointed out, many open questions concerning how and what different cortical structures do, and as Marsden and Obeso pointed out in 1994, we still do not completely understand how the basal ganglia operate. However, the findings of studies spanning more than a century that started with observations of the effects of the brain damage that leads to aphasia and subsequent studies of Parkinson disease, oxygen deprivation, electrophysiologic and lesion studies on animals, tracer studies of neural circuits, and ever more refined neuroimaging techniques demonstrate beyond reasonable doubt that these circuits play a key role in conferring human creativity — our ability to form novel concepts and choose among alternative courses of action.

Many other species are not automata and show aspects of creativity. Anyone who has had a dog knows that they learn new "tricks," and on their own can learn the meaning of the words that you speak and "read" your gestures. The "superdogs" studied by the Max Planck research group (for example, Kaminski et al., 2004) can rapidly learn the meaning of hundreds of words. Chimpanzees have devised and transmitted different tool-using cultures from one generation to another (Boesch, 1993). But the difference between human capabilities and those of other species is profound. Advances in genetics and an unlikely series of events have provided an opening in understanding how we came to be. In light of the fact that these circuits and their components do not materially differ in monkeys and humans, enhanced human neural capabilities must reflect increased "computational" efficiency—being better at performing the local operations that make us the unpredictable species. At least one genetic factor has been identified that does precisely that. There are clearly other genes, but a genetic anomaly in a large extended family and a series of carefully executed studies has provided a starting point in the path toward understanding why our brains differ from those of chimpanzees.

The FOXP2 Transcriptional Factor

The full name of the AAAS, the world's largest scientific organization, is the American Association for the Advancement of Science. "Advancement" has been the goal of science since its beginnings in the seventeenth century, but the path sometimes is improbable and meandering. In a session of the Boston Language Development Conference in 1998, I was listening to a moderately interesting paper, when Myrna Gopnick, a linguist at McGill University in Montreal, entered the room and startled me with an improbable statement. Myrna announced that she had news that would make both me and Steven Pinker happy. A few years earlier at another meeting in Kansas, I had debated the plausibility of Universal Grammar with Pinker, pointing out the problem posed by genetic variation.

Thousands of languages are spoken today; tens of thousands of languages that were once spoken have disappeared. The languages that we speak today have changed over time. English today isn't what Shakespeare spoke, nor is English the same in London, New York, Topeka, or New Delhi. As pointed out earlier, the Chomskian "language acquisition" model, which posits the existence of a Universal Grammar, in essence claims that children don't learn the particular rules of syntax or sounds of their native language. They instead activate principles and parameters off the shelves of their brain's hypothetical Universal Grammar, as though they were pulling items off the shelves of a very large language supermarket for their particular cart. But even in the biggest supermarket, some items may be sold out or discontinued, and you won't be able to get your favorite brand of jam, peanut butter, organic yogurt, or whatever.

The genetic variation that always exists in the population that defines a species would result in some items being missing off the shelves of the Universal Grammar's language supermarket. Some people are color blind or color deficient, owing to genetic variation. Some people have low metabolic rates and have to watch their diet. Some people are tall, some short, and so on. Therefore, if a genetically transmitted Universal Grammar really existed that stocked the total inventory of principles and parameters that yielded the syntactic rules, sounds, and syllable structures of every language on earth, some items would be missing and some children would never be able to acquire some of the specific items that their native language used. Specific syntactic rules of their native language would be missing.

In the sessions at the Kansas meeting, I had pointed out to Pinker that he could show that Universal Grammars existed, if he were able to find people with specific missing syntactic rules. If a Universal Grammar really existed, then some of the linguistic equivalents of the different brands of jam, peanut butter, and yogurt would be missing from some child's Universal Grammar.

Myrna Gopnick had been visiting London when she apparently turned on her television set and learned that some members of a large extended family had a similar pattern of speech and language problems that had a genetic basis. I have a videotape of a BBC science-news broadcast that shows some of the grandchildren of Hilda Kearney attempting to talk. British protocols concerning the anonymity of subjects apparently were quite different from those in force in the United States—Mrs. Kearney and her grandchildren were identified by name, and a street-map of Mrs. Kearney's London district was shown.

A genetically linked articulatory problem extended over three generations of Mrs. Kearney's descendants. The five children filmed were in a remedial program. One five-year-old girl had just learned to produce an unintelligible, distorted approximation of her first name.

English subtitles had been added because the children's speech was incomprehensible. One of the teachers also pointed out sentence comprehension and word-finding problems, but the school staff generally was upbeat about the children's intelligence and prospects. Marcus Pembry, a geneticist at London's Institute for Child Health, which was studying the children, also was on camera and pointed out the genetic locus of the children's problems. Not all of Hilda Kearney's children and grandchildren had these problems, which appeared to derive from a genetic anomaly that was not sex-linked.

Myrna Gopnick contacted the Kearney family and tested the ability of some of the afflicted members to comprehend two types of simple sentences. The sentences conveyed English plurals and past events. English can convey that you are referring to a plural noun by using a grammatical rule that adds the letter "s" to a noun—for example, as is the case for the "rule-governed, regular" noun pair "car-cars." In contrast, some noun pairs, such as "goose-geese" have "irregular" plural forms. The regular English past tense is formed by adding the letters "ed" to a verb. In contrast, the irregular past tense form of "see" is "saw." Linguists have been aware of these processes since the time of the Sanskrit grammarians thousands of years ago. Chomsky's great leap forward was to claim that the rules are innately activated by the Universal Grammar preloaded into everyone's brain. Chomsky's 1957 book *Syntactic Structures* proposed that we "generate" these sentences in our brain using "transformational rules" that convert an "underlying" mental representation in our mind to the sounds that we hear when we speak, or the written representation of the sentence, or for deaf persons, sign language. The "underlying" mental representation was supposed to reflect the manner by which the sentence's meaning was discerned. Chomsky has since modified his 1957 theory, but the basic premises have been preserved. In current versions of Chomsky's theory, an "I" language conveys semantic meaning, and the output "E" language is what you hear or read or for manual sign-languages; see Chomsky (2012).

The two sets of rule-generated English sentences that Myrna Gopnick presented to her subjects are among the few aspects of English that present a convincing case for Chomsky's argument.

They were the key examples of underlying mental representations conveying semantic information that were modified by "transformational rules" in Chomsky's lectures in the first class that he taught at MIT. They remain the examples cited in the first weeks of courses that teach Chomsky's views on language. However, when linguists attempt to describe the range of sentences that actually occur in a real language, they find that the basic premise of transformational grammar breaks down—the sentences fail to be rule-governed. In a conversation with Mark Johnson, a linguist who works on getting computers to comprehend sentences, Mark pointed out that as the number of different types of sentences increases, so do the number of rules—approaching the number of sentences. In other words, the sentences of a real language cannot be described using the algorithms commonly used by linguists following Chomsky's lead. That was the conclusion reached years ago by Marcel Gross in his 1979 paper "On the Failure of Generative Grammar."

The Syntax Deficits of Family KE

Myrna Gopnick was ecstatic because she had found that the afflicted members of family KE (that's the nomenclature presently used to describe Hilda Kearney's extended family) had difficulty understanding sentences that used the regular forms of the English past tense and plural. Her 1970 paper published in the journal *Nature* was interpreted as proof for a "language gene" that somehow coded two specific syntactic rules of the Universal Grammar. However, there was a glaring hole in her study. Gopnick's study inherently could not demonstrate any such specific deficits because she had not tested the KE family's ability to comprehend any other aspects of English syntax. Did the members of the KE family have general problems comprehending syntax? Did they have other cognitive problems extending beyond syntax or comprehending the meaning of sentences? Why were they almost unable to talk? Steven Pinker nonetheless organized a session featuring the putative language gene at the annual meeting of the AAAS. You can read about the putative language gene in Pinker's 1994 book *The Language In-*

stinct. However, though further research showed that there is a gene responsible for the speech and sentence comprehension difficulties of Hilda Kearney's descendants, it's neither a language or a speech gene. And this gene, FOXP2$_{human}$, opened a new pathway toward understanding why we can act and think as we do.

The Institute for Child Health (ICH) in London had been systematically studying the extended KE family. The ICH studies revealed a syndrome—a suite of severe movement disorders involving lip and tongue movements, cognitive deficits, and a general problem in comprehending distinctions in meaning conveyed by syntax. Faraneh Vargha-Khadem and her colleagues (1995) tested 21 members of the KE family. Thirteen had the anomalous autosomal gene; eight were unaffected. With the exception of one person, the subjects were the same as those tested by Gopnick. The sentence comprehension problems that occurred in the afflicted members of family KE were general, not limited to two bits and pieces of the hypothetical Universal Grammar; the sentences that could be "generated" by two "transformational rules." On tests of comprehension of grammar, the affected members of the KE family were significantly impaired comprehending sentences that had relative clauses, repeating words, changing the letter of a word to form a different word. Moreover, they clearly understood the rule that produces regular English plurals and tended to overgeneralize it. They had as much difficulty producing irregular past tense and plurals as regular forms.

They could not simultaneously stick out their tongues while pursing their lips, reflecting difficulty coordinating these simple motor gestures. They could not repeat a two-word sequence though they tried again and again. They had difficulties repeating nonwords. They had problems choosing the correct word when they attempted to talk (Lai et al., 2001; Vargha-Khadem et al., 1998, 2005; Watkins et al., 2002). Cognitive tests, such as the Wisconsin Card Sorting Test or Odd-Man-Out test that would have targeted their ability to change the direction of a thought process were never administered. However, on standardized Wechsler Intelligence tests, affected members of family KE had significantly lower scores than their nonaffected siblings, a result that rules out the presence of environmental factors influencing the results.

Some of the afflicted individuals had higher nonverbal IQ scores than unaffected members, which has led to some studies labeling FOXP2 a "speech and language" gene. However, there was a highly significant difference in mean test scores: the mean for the affected members was 86 (a range of 71–111) versus a mean of 104 (a range of 84 to 119) for unaffected family members. Moreover, verbal working memory was significantly impaired on digit span tests (repeating lists of numbers forward and backward)—an aspect of cognitive executive control that, as noted earlier, involves prefrontal cortical activity linked to posterior cortex in basal-ganglia circuits.

A similar pattern of deficits, though less severe, has been found in children identified at Hasbro Children's Hospital in Providence as having verbal apraxia. Many of these children suffered ischemic (reduced bloodflow) oxygen deficits during birth, which can degrade the basal ganglia (Robertson et. al., 1999; Jeong et al., 2002; Kuoppamaki et al., 2002; Chie et al., 2004). Speech deficits similar to those seen in the KE family occur. Multisyllabic words and consonant-vowel-consonant (CVC) syllables reduce to a consonant-vowel (CV) syllable, sometimes to a CV followed by a long pause and another CV sequence. A monosyllabic CVC word such as "cat" turns into the two syllables "ca + ta" or simply "ca." Speech slows down, vowels are prolonged, long pauses occur. Sentence comprehension deteriorates owing to difficulties interpreting distinctions in meaning conveyed by syntax. Cognitive capabilities such as changing the direction of a thought process that involve cortical-basal-ganglia circuit activity deteriorate.

In Parkinson disease, it is almost a certainty that cognitive flexibility will be impaired when sentence comprehension is impaired (Hochstadt et al., 2006). The woman studied by Pickett et al. (1998) who had bilateral lesions in the caudate nucleus and putamen of the basal ganglia had severe difficulties comprehending sentences that had even moderately complex syntax. She had profound difficulties on the Odd-Man-Out test, which assesses cognitive flexibility. That pattern also occurred in hypoxic Everest climbers. In short, the affected members of the KE family had the suite of motor, cognitive, and linguistic impairments seen when the basal ganglia are compromised.

Science isn't always a well-ordered, dispassionate enterprise. The editors of the internationally acclaimed journal *Nature* at first refused to accept a paper by disputing Gopnick's findings that Faraneh Vargha-Khadem and her colleagues submitted. The reason advanced by *Nature*'s editors was that Gopnick had published a short paper, so the publication of a detailed account of the actual sentence comprehension and nonlinguistic cognitive problems that existed in this family was unwarranted. Fortunately, *Nature* is not the only scientific journal in world. The Vargha-Khadem et al. (1995) paper was published in the *Proceedings of the National Academy of Sciences of the US (PNAS)*. Subsequent papers by the ICH group and collaborating scientists were published in the *Nature* group of journals.

What Anomalous FOXP2 Did to the KE Brain

MRI imaging of affected KE family members confirmed basal ganglia anomalies. MRIs showed that the caudate nucleus of the basal ganglia is abnormally small bilaterally. The caudate nucleus, as noted earlier, is active in shifting the direction of a thought process (for example, Monchi et al., 2001, 2006), and in a wide range of cognitive tasks including mental arithmetic, sentence comprehension, and retrieving words from the brain's dictionary (Kotz et al., 2003; Rissman et al., 2003; Wang et al., 2005). The reduction in caudate nucleus volume was "significantly correlated with family members' performance on a test of oral praxis, nonword repetition, and the coding subtest of the Wechsler Intelligence Scale" (Watkins et al., 2002). Watkins and her colleagues concluded that these "verbal and nonverbal deficits arise from a common impairment in the ability to sequence movement or in procedural learning."

The putamen and globus pallidus of the basal ganglia, which also are active during these cognitive tasks, were abnormal unilaterally in affected members of the KE family, as were the angular gyrus, cingulate cortex, and Broca's area. Functional abnormalities were found in these and other motor-related areas of the frontal lobe (Vargha-Khadem et al., 1995, 2005). PET imaging showed overactivation of the caudate nucleus in two afflicted individuals during a simple word repetition task (Vargha-Khadem et al., 1998). fMRI studies that compare afflicted members of the KE family with both their "normal" siblings and age-matched controls show that underactivation occurs in the putamen, Broca's area, and its right homologue (Liegeois et al., 2003), a finding that is consistent with circuits connecting the striatum and Broca's area (Lehericy et al., 2004). The affected neural structures thus include those supporting cortical-striatal-cortical circuits implicated in motor control, including talking, various linguistic tasks, and the range of cognitive activities that involve executive control that have been discussed in the previous chapters.

Transcriptional Factors and Human Evolution

An egg is an egg, but a cook can boil it, beat it with a fork and fry it to make an omelet, or stir it into the batter for a cake. The cook is the master of the egg's fate, determining the form in which the egg reaches the table. Transcriptional factors are "master" genes that play a major role, like that of the cook, in determining how other genes manifest themselves—how they shape a living organism. Transcriptional genes, for example, are responsible for our backbones having 5 instead of 3 lumbar vertebrae, or our brains having increased synaptic plasticity so as to enhance human cognitive ability. Genetic studies of the KE family led Simon Fisher of Oxford University to identify the FOXP2 transcriptional factor that plays a central role in why we can act and think in a manner quite distinct from chimpanzees, though we share almost 99 percent of our genes (Fisher et al., 1998). The cognitive and motor deficits and neural anomalies observed in the afflicted members of the KE family resulted from their having only one copy of the human version, $FOXP2_{human}$, of this transcriptional factor. This deficiency affected the way that their genes expressed themselves. In light of the KE family studies, the reader might assume that FOXP2 acts only on the brain. But that really isn't the case.

The information in the double helix of DNA that constitutes the genetic code has to be transcribed into a different form, single-stranded mRNA that later is translated into functional proteins and the structures of living organisms. Transcription factors are genes that govern this process—proteins that bind to particular DNA sequences near a gene that they regulate so as to control the degree to which they release information to the mRNA. In effect, transcription factors are genes that determine how other genes ultimately result in the structures that constitute a living organism and how they function. FOXP2 is one of many transcription factors that exist in all mammals, birds, and other creatures. The DNA sequence of the avian form, Foxp2 (the lowercase spelling indicates that it's not the human version), is 98 percent identical to the human version. The mouse form of Foxp2 controls the embryonic development of the lungs, the intestinal system, the heart and other muscles, as well as the spinal column of mice (Shu et al., 2001).

But the focus here is on the brain. Cecilia Lai and her colleagues in their 2003 study compared the expression of the mouse version, Foxp2, and human version, FOXP2$_{human}$, during early brain development in both humans and mice (Lai et al., 2003). We are separated from mice by 75 million years of evolution (Mouse Genome Sequencing Consortium, 2002). The Foxp2 and FOXP2$_{human}$ genes encode a forkhead transcription factor, a protein that regulates the expression of other genes during processes that mark embryonic development, such as signal transduction, cellular differentiation, and pattern formation. Mutations to other forkhead transcription factor genes have been implicated in a number of developmental disorders. In the case of family KE, the mutation changes a single amino acid in the DNA-binding region of the protein, and that single change apparently leads to protein dysfunction. The areas of expression of FOXP2$_{human}$ and Foxp2 in both the human and mouse brain are similar and include the neural structures that form the human cortical-striatal-cortical circuits involved in motor control and cognition—the thalamus, caudate nucleus, and putamen, as well as the inferior olives and cerebellum (other subcortical structures). These neural structures are all intricately interconnected. The cerebellum, which receives inputs from the inferior olives, is involved in motor coordination. The cortical plate (layer 6), the input level of the cortex, is also affected by the FOXP2 mutation.

The subsequent focus on the role of FOXP2$_{human}$ in human evolution follows from its being one of the few genes that has been shown to differ from its chimpanzee version (Chimpanzee Sequencing and Analysis Consortium, 2005). A "human" version, FOXP2$_{human}$ evolved that differs from the version found in chimpanzees, Foxp2chimp, during the 6 or 7 million year period of evolution that separates humans and chimpanzees. In that period, FOXP2$_{human}$ underwent two substitutions in its DNA sequences, causing two amino acid changes in FOXP2 protein compared to one amino acid substitution between chimpanzees and mice over 70 million years. And an additional change, that still is the subject of ongoing research, seems to have occurred in the period in which anatomically modern human beings appeared some 260,000 years ago.

The date of the "selective sweep," approximately 260,000 years ago, which resulted in FOXP2$_{human}$ being present throughout the human population, was established by Wolfgang Enard and his colleagues in their 2002 paper. The research was carried out by Svante Paabo's research group in 2002 at the Max Planck Institute for Evolutionary Anthropology in Leipzig, Germany. As the previous chapter pointed out, selective sweeps occur when a gene results in a truly significant advantage in the Darwinian "struggle for existence"—an individual's having more surviving children. Selective sweeps in different human populations have occurred for adult lactose tolerance, living at extreme altitudes, living in arid regions, and so on. In most instances, it is unclear what genes do when we find a difference between the chimpanzee and human versions of a gene. However, in this instance the effects of a FOXP2$_{human}$ anomaly in the KE family showed that it playing a key role in quintessentially human attributes—speech, language, and cognition.

GMO Mice

Further studies at the Max Planck Institute and UCLA (the research project has become a worldwide endeavor) show that the human version of FOXP2 basically ramps up information transfer and associative learning in the basal ganglia. In light of basal ganglia activity in both associative learning and motor control, this would account for afflicted members of the KE family being unable to learn and execute the complex motor acts that enable us to talk, being unable to execute other internally guided motor acts, and their cognitive deficits. It had been uncertain whether the human version of FOXP2 would make a difference in how brains work because the anomalous version of FOXP2 in family KE and other individuals does not correspond to either normal chimpanzee Foxp2 or human FOXP2$_{human}$. The Max Planck GMO mouse study and an independent UCLA study resolved this question.

At a meeting in San Diego in 2006, Svante Paabo, the Max Planck team leader, told me about his research group's intention to "knock in" the human version of FOXP2— FOXP2$_{human}$—into mice, replacing the normal "wild" mouse version of Foxp2. Svante jokingly asked whether I thought that the mice would then be able to talk. My reply, in a similar spirit, was, "no." There is more to the neural bases of talking than the local operations of the basal ganglia. The mouse brain lacks the cortical areas that project to the basal ganglia in humans, and mouse cortex clearly differs in other ways from that of humans. More to the point, it's almost a certainty that other genes were involved in the evolutionary processes that conferred human motor and cognitive ability. But we now have a start in understanding why the human cortical-basal ganglia circuits allow us to act and think in a manner that no chimpanzee could ever attain.

Using techniques analogous to those used to produce GMOs, genetically modified wheat, corn, tomatoes, chickens, and so on, the Paabo laboratory "knocked in" the human form of FOXP2 into mice (Enard et al., 2009; Reimers-Kipping et al., 2011). The DNA of the mouse version of Foxp2 is three amino acid substitutions away from the human version and also has other differences. When FOXP2$_{human}$ was knocked into mouse pups, their vocal calls were somewhat different than the calls of mouse pups that had the normal "wild" version of Foxp2. When the wild form of Foxp2 was knocked out in mouse pups, they died soon after, reflecting its role in lung and cardiovascular development.

However, the most significant difference was that apparent in the cortical-basal ganglia circuits that we have been discussing—the circuits that regulate motor learning and motor control and confer cognitive flexibility. Increased synaptic plasticity occurred in basal ganglia neurons, as well as increased dendritic lengths in the basal ganglia, thalamus, and layer 6 of the cortex. In particular, $FOXP2_{human}$ increased synaptic plasticity in medium spiny neurons in the basal ganglia and in the substantia nigra, another structure in cortical-basal circuits (Alexander et al., 1986; Cummings, 1993). Increasing synaptic plasticity has been shown to enhance associative motor learning in mice (Jin and Costa, 2010). The effect of increased synaptic plasticity in enhancing associative learning was expected. As the first chapter of this book pointed out, Hebb in 1949 formulated the theory that has since guided research on the basic operations of the brain. Synapses transfer information from one neuron to another. Synaptic modification is the neural mechanism by which the relations that hold between seemingly unrelated phenomena are learned. The process by which we learn anything—motor acts, names, concepts, and so on—involves modifying synaptic "weights"—the degree to which synapses transmit information to a neuron. That's the case for creatures as far removed on the evolutionary scale as humans and mollusks (Carew et al., 1981).

The increased synaptic plasticity and connectivity in the medium spiny neurons of the basal ganglia that resulted from $FOXP2_{human}$ is especially significant in light of their role in associative learning. Other recent and ongoing independent research projects (discussed in chapter 2) show that these neurons and dopamine-activated neurons in the basal ganglia and prefrontal cortex in essence guide associative learning by coding the expectation of achieving a desired goal or avoiding aversive outcomes (Bar-Gad and Bergman, 2001; Joshua et al., 2008; Assad and Eskandar, 2011). In short, information stored in the synaptic weights of these neurons directs the process of associative learning, allowing animals to learn to perform complex linked sequences. In humans, similar processes would account for our learning complex grammatical "rules," as well as the equally complex rules that guide our interactions with other people, other species, and the conditions of daily life, and our responses to new and novel situations.

Studies ranging from electrophysiologic recordings of neuronal activity in the basal ganglia of mice and other animals as they learn tasks (Graybiel, 1995; Mirenowicz and Schultz, 1996; Joshua et al.; 2008; Jin and Costa, 2010; Assad and Eskandar, 2011) to studies of Parkinson disease patients (Lang et al., 1992; Monchi et al., 2001, 2006a, 2006b, 2007) and birds (Brainard and Doupe, 2000) confirm the critical role of the basal ganglia in associative learning, planning, and executing motor acts, including motor control for vocal communication.

Supercharged Neural Circuitry?

As I've pointed out repeatedly, our knowledge of how brains work is imperfect. However, the chain of serendipitous events that led to the discovery and behavioral and neural effects of the $FOXP2_{human}$ transcriptional factor has provided an opening into why humans can learn and execute the complex motor acts that are involved in talking, and the neural bases that make us so unpredictable and creative. We are only beginning to understand how transcriptional factors shaped the human brain. Ongoing research projects show that FOXP2 is not the only gene that enhanced human cognitive ability. The almost simultaneous publication from Daniel Geschwind's laboratory at UCLA confirmed the role of the human version of FOXP2; it also pointed to other genes enhancing information transfer in the human brain. Using a different technique than the Max Planck research group, Konopka et al. (2009) confirmed that FOXP2 "upregulated" genes that, in turn, affect the caudate nucleus—the basal ganglia structure that is the neural engine that confers creativity though its role in allowing us to change the direction of a thought process. The UCLA group also found 61 other human genes whose expression was unregulated by the human form of FOXP2. Five of these genes, expressed in the brain, were under positive selection in the human lineage. The data suggest to Konopka and her colleagues that, "a subset of differential FOXP2 targets may have co-evolved to regulate pathways involved in higher cognitive functions." Tod Preuss and his colleagues in their 2004 *Nature* genetics review paper looked at the several hundred genes that differ in humans and chimpanzees. They view the hundred or so different genes as a lower limit. Their paper pointed out that virtually all of the genes expressed in the human brain that are known to differ between chimpanzees and humans are unregulated. In another study, Green et al. (2010) identified "highly accelerated regions" (HARs) of the human genome that differ from the Neanderthal genome. These genes appear to be associated with cognition. Their "cost" is that they seem to be associated with schizophrenia, autism, and other mental illnesses.

The "mad" genius stereotype may have some truth in it. In the course of evolution, we humans may have acquired unstable genes that enhance cognition, trading instability for a chance at genius. Pal et al. (2010), for example, isolated an aberrant gene that occurs in people who have epileptic seizures. My small contribution to this study was speech analysis that showed that the aberrant gene also results in speech production deficits similar to Parkinson disease, but it isn't clear whether it has any bearing on the evolution of the human brain. The ASPM gene (Zhang, 2003) might be involved in the enlargement of human brain size (an aberrant version results in small brains).

We have a long way to go before we can identify all of the genes that shaped the human brain. Current research is at the starting point of understanding of how transcriptional genes may have ramped up our neural pathways. However, the $FOXP2_{human}$ studies that have integrated genetic, neurophysiologic, and behavioral evidence suggest that the fundamental distinction between the human and nonhuman primate brain is one of computational efficiency. That distinction may account for last stages of the evolution of the hominin brain.

Who Else May Have Had Semi-Supercharged Brains?

Neanderthals have posed a problem to everyone studying human evolution since the first Neanderthal skull was discovered in 1856 in the Neander valley in Germany. (The name Neanderthal signifies

"Neander valley" in German.) Various theories have been proposed concerning Neanderthals. Some anthropologists consider them to be a human variety, but the weight of evidence points to their being a different species. This didn't preclude their mating with humans when some of our distant ancestors arrived in Eurasia about 40,000 years ago. As Darwin pointed out, the concept of species necessarily becomes elastic when it becomes evident that one species can gradually evolve into a different species. We now know that modern humans and Neanderthals mated in the period in which humans first arrived in Eurasia. Svante Paabo's research group compared their draft sequence of the Neanderthal genome that was derived from three Neanderthal bones unearthed in Croatia with contemporary humans from Africa and elsewhere (Green et al., 2910). About 1 to 4 percent of the genes in non-African humans are estimated to result from Neanderthal-human mating.

These comparisons of the Neanderthal genome with that of humans and the genome of a second extinct hominin species, the Denisovans, show that more work is necessary to fully specify the nature of $FOXP2_{human}$. The Denisovans, whose fossil remains were recently found in Siberia, are more closely related to Neanderthals than to us, probably splitting off from the Neanderthals about 200,000 years ago (Reich et al., 2010). The critical fact is that the genetic distinction that was thought to completely distinguish the human form from chimpanzee Foxp2, the two amino acid substitutions, has been found in Neanderthal fossils (Krause et al., 2007) as well as in a Denisovan fossil (Meyer et al., 2012).

If the completely human version of FOXP2 only involves the two amino acid substitutions, then its presence in Neanderthals and Denisovans would push back the date of the selective sweep on this gene to the last common ancestor of Neanderthals to between 370,000 and 450,000 years ago (Green et al., 2010). That is implausible, given the evidence that established the date of the selective sweep that spread $FOXP2_{human}$ throughout the human population. Wolfgang Enard and the Paabo group's research team dated the selective sweep to 260,000 years in the past, well after the divergence of humans from either Denisovans or Neanderthals.

There is almost no variation in the normal form of FOXP2$_{human}$ in present-day human populations. Graham Coop and his colleagues at the University of Chicago first pointed out that this precludes the selective sweep on FOXP2$_{human}$ occurring 450,000 years ago. The Coop et al. (2008) study noted that there would be more variation in FOXP2$_{human}$, than the Enard study found because mutations accumulate over the course of time. We all seem normally to have the same form of FOXP2$_{human}$.

Denisovans, like Neanderhals, mated with humans, the Denisovans with the ancestors of present-day Melanesians, so it is possible that Neanderthals and Denisovans having a form of FOXP2 two amino acid substitutions removed from chimpanzees may have resulted from mating with modern humans, but the geographical dispersion makes this highly improbable. Moreover, Meyer and his colleagues in their 2012 study found other highly significant genetic differences between human and Denisovans that we will return to.

Svante Paabo, his colleagues, and other geneticists are convinced that an additional genetic factor distinguishes FOXP2$_{human}$ from both the Neanderthal and Denisovan versions; a mutation specific to humans occurred on FOXP2, advantageous in the Darwinian struggle for existence that accounted for the selective sweep that occurred 260,000 years ago. The genetic event that drove this selective sweep remains unclear, but the Paabo group is focusing in on the region of the DNA sequence (exon 7) of the FOXP2 transcriptional factor. You can refer to the Ptak et al. (2009) research report for the details.

In short, Svante Paabo and his colleagues, through a series of brilliantly conceived and executed studies, have established how FOXP2$_{human}$ supercharges the cortical-basal ganglia circuits that confer the human cognitive capacities that make us both so creative and unpredictable. The GMO mouse knock-in studies provided the answer. They also established the timing of the selective sweep of FOXP2$_{human}$, but the full, detailed, genetic differences that distinguish FOXP2$_{human}$ from Neanderthal, Denisovan, and chimpanzee versions remains the subject of continued research. We still don't know the complete genetic recipe for FOXP2$_{human}$.

The End of the Neanderthals

The extinction of the Neanderthals and Denisovans most likely follows from the default hypothesis, supported by the historical record—Charles Darwin's gloomy view of how different humans often interact when they first meet. In his notebooks, Darwin wrote,

> When two races of men meet they act precisely like two species of animals,—they fight, eat each other, bring diseases to each other &c., but then comes the more deadly struggle, namely which have the best fitted organizations, or instincts (ie intellect in man) to gain the day . . . man acts on & is acted on by the organic and inorganic agents of this earth like every other animal (Browne, 1995, p. 399).

Human groups probably behaved more or less in the same manner toward Neanderthals and Denisovans as they did to "other groups" in the recent past and often do today. The historical record unfortunately is full of examples of colonists who have better weapons or tactics, wiping out an indigenous people. Darwin's account of Argentina in the early nineteenth century is only one example. It is apparent that Neanderthals didn't survive for very long after the first humans arrived in Eurasia. Revised, better dating of Neanderthal sites and fossils shows that Neanderthals became extinct at about the 40,000 years ago date at which humans appeared on the scene (Pinhasi et al., 2011). In the next chapter, we will return to examine Neanderthal culture as a marker for signs of creativity, or its absence.

Birds and Us

Whereas primates, other than humans, don't seem to be able to learn new vocalizations, some species of birds do that all the time. Mynah birds and parrots probably are the best known imitators, but other species are quick to learn new songs. Birds don't have a cortex, but they have basal ganglia circuits similar to those of mammals.

Doupe and her colleagues (2005) showed that the neural bases for vocal imitation in birds is a "basal ganglia forebrain pathway."

The bird basal ganglia homologue, area X, is modulated by the avian form of Foxp2. Zebra finches learn a new set of songs each year, making them ideal candidates for studying the neural mechanisms involved in learning new songs. Avian Foxp2 is at a higher level in area X during the period when Zebra finches learn their song repertoire. In short, vocal learning and vocal control in humans and birds involve neural circuits that include the basal ganglia.

The Foxp2 bird studies also show that genes, even transcriptional genes, don't work in a vacuum. It's clear that human language and cognitive ability do not develop when a child is isolated from normal human contact for years (Curtiss, 1977), but no one has yet shown the direct effects of environment on human gene activity that might be directly implicated in language. It turns out that the maximum level of Foxp2 expression in area X of the bird brain occurs when a male is courting a female, reaching a higher level than when he is alone, practicing his new songs (Teramitsu and White, 2006). The bird's "social" context acts on the release of the transcriptional gene. Are similar phenomena at work in mammals, including us?

The Evolution and Utility of Cars

I began the discussion of the functional organization of the brain by comparing cars and brains. Automotive design again provides a parallel. The high-performance engine in the racing version of the car that my wife and I drive doesn't really differ from the fuel-efficient engine in our car, except that it's supercharged. Yokohama Heavy Engineering (aka Subaru) added a supercharger, a device that pushes more air into the cylinders, in their high-power WRX and STI Impreza engines, which aren't very different from our car's engine except for the supercharger. The add-on supercharger boosts acceleration and speed to race-car levels, about 150 mph. We don't yet know much about the other transcriptional factor add-ons that may have been made to the chimpanzee-like brains of early hominins, but $FOXP2_{human}$ supercharges our cortical to basal ganglia circuits. The hundred or so genes that differ in chimpanzees and humans, particularly those that

117

enhance cell-to-cell transmission, may be supercharging cortical-to-cortical circuits and the manner in which the basic neuronal computing elements of our brain function.

The evidence that I have reviewed argues against our having domain-specific neural language organs, mathematical organs, and so on. So it is with a car. It's obvious that your car can take you to the supermarket, school, Yellowstone National Park, or home. It is not domain-specific in the sense that a particular car can be driven only to the supermarket. You don't need another car to drive to school. The same car potentially can take you to any location that can be reached on a road. But cars are useful only if roads exist. Roads are a necessary factor. Roads existed before cars, and people could travel using horse-drawn wagons and carriages, riding a horse or donkey, pedaling on a bicycle, or simply walking. But cars opened new possibilities. Everyone wanted a car. Cars changed the way we live and changed the very shape of the human landscape.

The increased computational power of the human brain is in some ways similar to the distinction between cars and the mode of travel in pre-car days. The horseless carriage was a leap forward that derived from adding an engine to a carriage, but it would have been a useless toy if roads had not already existed. The studies discussed in this chapter suggest that transcriptional genes, upregulating neural processes, created a supercharged human brain. But if brain size had not already increased, allowing hominins to acquire and access a vast store of knowledge, concepts, and plans of action, our supercharged brain would not have yielded the cultural revolution that started 200,000 or 300,000 years ago. Brain size stands in the same relation to human cognitive, creative capacities as roads did to making a horseless carriage an instrument of profound change.

The State of the Art

Do we fully understand why human creative capacities exceed those exhibited by our closest living relative—the chimpanzee? Do we know whether enhanced human cognitive capacity was the factor that resulted in our distant ancestors replacing Neanderthals? And if that was the case, what genetic events may have enhanced human creative capacity relative to Neanderthals? Remarkable progress has been made in understanding the evolution and nature of the human brain. However, we are only starting to understand the interactions between genes, neural development, and—in plain English—how brains work. The FOXP2 transcriptional factor is clearly involved in those processes, but the precise character of $FOXP2_{human}$ is still to be determined. It is apparent that other genes differentiate human brains from those of archaic hominins such as the Neanderthals, and Denisovans.

In the words of Ann Gibbons, one of *Science Magazine's* writers specializing on human evolution, Mathias Meyer and his co-workers achieved a "stunning technical feat." The detailed genetic map from a fragment of a finger bone of a Denisovan girl whom they estimate lived 80,000 years ago, "as sharp a picture of this ancient genome as a living person's," and "produced a 'near complete' catalog of the small number of genetic changes that make us different than the Denisovans, who were close relatives of the Neanderthals." (Gibbons, 2012, p. 1028).

The conclusion reached by Meyer and his colleagues regarding the difference between our brains and those of the Denisovans again involves the FOXP2 transcription factor that increase synaptic plasticity in the human brain's basal ganglia, but other genes may have enhanced the cognitive capacities of the human brain. The team of scientists included the Paabo group, Harvard Medical School, and research groups ranging from Arizona to Beijing and Novosibirsk, Siberia. The detailed reconstruction of the genome sequence of the Denisovan girl was compared with the genomes of eleven present-day humans from different ethnic groups and other human genetic data bases (Meyer et al., 2012). The analysis identified Denisovan genes that affect the skin, eyes, and teeth. Denoisovans, like Neanderthals, apparently had large molar teeth. Eight Denisovan genes that are involved with the basic computational architecture of the brain (the connections and information transfer between neurons) differed from human genes. As Meyer and his colleagues note (the genes are identified by the letter and number sequences):

> Four are associated with axonal or dendritic growth (SLITRK1, KATNA1) and synaptic transmission (ARHGAP32, HTR2B) and two have been implicated in autism (ADSL, CBTNAP2). CNTNAP2 is also associated with susceptibility to language disorders and is particularly noteworthy as it is one of the few genes known to be regulated by FOXP2, a transcription factor involved in language and speech development as well as synaptic plasticity. It thus is tempting to speculate that the crucial aspects of synaptic transmission may have changed in modern humans. (Meyer et al., 2012, pp. 4–5)

In short, at least one human gene, that $FOXP2_{human}$ acts on, differs from its Denisovan version. Since Denisovans and Neanderthals shared a common ancestor who diverged from our human ancestors between 70,000 and 700,000 years ago, there may be a similar critical difference between the Neanderthal and human versions of CNTNAP2. A detailed Neanderthal genomic map using Mathias Meyer's new technique may reveal whether this is the case, and whether other genetic distinctions that enhanced synaptic plasticity and connectivity account for our supercharged brains.

There are still more open questions than answers, and research institutes focusing on these issues are being founded throughout the world. One example—the Salk Institute in California is establishing a Center for Integrative Biology that will focus on understanding the interactions between genes and how neurons and neural circuits signal information. Eminent scientists such as Terry Sejnowksi, who have already made their mark in neuroscience, will be examining an aspect of the DNA of neurons (methylome) that affects the development of the genes involved. The planned study also may further our understanding of the nature of human mental disorders. It's an exciting time and I wonder where we will be in ten years. But it is clear that we humans have the neural capacity for being unpredictable.

Chapter Five

Stones, Bones, and Brains

A s the studies reviewed in chapter 4 showed, the FOXP2 gene probably reached its human form about 260,000 years ago. But other transcriptional factors, which were touched upon in the previous chapter, most likely were involved in shaping the neural bases of human creativity. Can we bring other information to bear on the question of when and where the neural capacity for human creativity evolved?

The Archaeological Record

Some of the evidence is in the ground beneath you if you happen to be in the right place. If you were to rummage through someone's home, the books, pictures, kitchen utensils, and clothes would form a picture of how they live, and might suggest whether they were creative persons you might want to know. The archaeological record is a similar resource, but it's usually the case that few artifacts survive, and we often know almost nothing about the conceptual space of the people who made or used the objects that have been unearthed. Moreover, the archaeological record preserves artifacts, and artifacts always reflect the values of a culture. We, therefore, always must attempt to take into account the role of culture and cultural aggregation before equating artifacts with cognitive capacities. For example, it is clear that people living today in extreme poverty are unlikely to be able to exercise their full creative capacity. Some cultures actively inhibit creativity. If you were able to travel back in time to 1860, would it be apparent that the descendants of some Japanese rice farmers would be designing hybrid cars? Or that they and the great-grandchildren of unkempt European peasants would be designing, building, and routinely using CT scanners, Airbuses, or Boeing 747s?

The reverse sometimes occurs. When the Royal Navy's exploration vessel *Beagle* reached Tierra del Fuego in 1832, three indigenous Yaghans were on board, as well as Charles Darwin. The three Yaghans, York Minster, Fuegia Basket, and Jemmy Button, were returning home. Their silly names had been coined three years earlier by the *Beagle*'s crew when they were transported to England on the *Beagle*'s previous voyage. During their event-packed three years in Britain, the Yaghans had all learned English to varying degrees of fluency and were neatly dressed in proper British clothing. On board the *Beagle*, they ate with forks, spoons, and knives and were familiar with the tools, conveyances, and many of the customs of early nineteenth-century England.

Darwin was astonished to see their kinsmen—stark naked except for ratty animal skins flung over their shoulders. The Yaghans on shore were using primitive stone tools, similar to ones first made by *Homo habilis* millions of years ago in Africa. However, the Yaghans were modern humans whose ancestors had migrated down from the South American mainland to the islands of Tierra del Fuego. And none of the indigenous inhabitants of South America ran about naked. The complex Inca civilization had flourished there. The Incas had, before the arrival of the Spanish Conquistadors, smelted gold, produced intricate metalwork and weavings, and built major cities and roads through the Andes. In some manner, the Yaghans had lost, or perhaps never attained, that technical base. However, Darwin was rightly convinced that they were members of his own human species.

Culture can release cognitive capacities, even for apes. Alan and Trixie Gardner (1969) raised chimpanzees from birth in a setting where people communicated with them and each other using American Sign Language (ASL). The Gardner chimpanzees communicated their wants and thoughts using about 150 words. The chimpanzees had progressed beyond the stage of thinking of words as though they had fixed referents. Their words' semantic referents instead reflected their life experiences. The ASL word "meat," for example, referred to the hamburger on the table as well as to the shrink-wrapped packages in the supermarkets to which the young chimpanzees were taken on shopping trips. Washoe and the other Gardner chimpanzees signed to themselves as they "read" picture books and related, albeit simply, past events. Savage-Rumbaugh replicated the Gardner studies with the bonobo (pygmy chimpanzee) Kanzi, using a manually operated keyboard that controlled a speech synthesizer (Savage-Rumbaugh and Rumbaugh, 1993). Kanzi responded correctly to spoken English sentences, even when he was asked to perform very odd tasks, such as putting a ball into a refrigerator. In the filmed record of that event, Kanzi looked quizzically at Savage-Rumbaugh as he opened the refrigerator. No one in 1969 had thought that chimpanzees were capable of using words and comprehending simple syntax. No one in 1800 would have thought that you and I would be able to routinely hurtle past each other driving cars at closing speeds exceeding 120 an hour, narrowly missing each other on either side of a painted white line.

It also is apparent that the state-of-the-art technology of the people living in any given place or time does not in any absolute sense reflect their cognitive, or even their motor capacities. Benjamin Franklin rode in a coach; you, the reader, can ride in a jet airplane, but you're not necessarily wiser. Nonetheless, the archaeological record can provide some inferences concerning the cognitive capacities of hominins who lived millions of years in the past if we take into account two factors: (1) the baseline provided by our close nonhuman primate relations, and (2) the rate of change.

Chimpanzees again provide the baseline. Although present-day chimpanzees are not living examples of the common ancestor of humans and apes, we can be certain that any artifact that a chimpanzee can make or use could have been fabricated or used by our common ancestor. Jane Goodall in her 1986 book *The Chimpanzees of Gombe* showed that chimpanzees can fashion and use simple tools and opportunistically adopt found objects. The Gombe chimpanzees, for example, crafted termite-sticks by stripping twigs off branches so that the sticks could be used to fish out termites when they were inserted into mounds. The result was a sort of termite lollipop. Leaves were mashed into "sponges" to soak up and drink water. Older chimpanzees transmitted these technologies to young chimpanzees. Other examples, as well as the complex social structure of the Gombe chimpanzees, are described in Goodall's superb book. Chimpanzee groups isolated by geography developed their own tool technologies. The Tai forest chimpanzees used club-like pieces of wood or small stones to hammer open nuts (Boesch, 1993). Hard-shelled nuts had to be cracked open with stones. The chimpanzees had beforehand searched for suitable stones and stored these stones for the next nut-cracking season. However, no chimpanzee has mastered the art of modifying stones to make cutting tools.

Static Periods

When the archaeological record of the first five million years or so of hominin evolution is examined, there is no evidence of any attempt to make stone tools. The toolkit of the hominins who lived in that distant time probably was similar to that of present-day chimpanzees—branches and leaves They may have used animal bones and found stones as hammers and perhaps worked other material that has since decayed. They also may have hurled stones and sticks while hunting, though chimpanzees are hardly ever able to hit their targets. The first "solid" evidence, literally writ in stone, for technology more advanced than that of a chimpanzee dates back to between two and three million years ago. Louis Leakey, his wife, Mary, and their colleagues in 1964 attributed

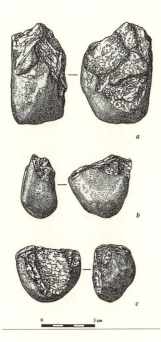

Figure 5.1. Oldowan tools. The earliest stone tools dating back two or three million years are the first "solid" evidence for tools that present-day chimpanzees don't make. They may have been made by the hominin species *Homo habilis*. They can readily be mistaken for stones chipped by natural events. Source: Lawrence Barham.

the chipped stone Oldowan tools (some are sketched in figure 5.1) to the fossil hominims whom Louis Leakey named *Homo habilis*. The fossil record is very sparse between 3 and 2 million years ago, but the oldest dated Oldowan tool is dated to about 2.7 million years ago. There is no evidence for technology more advanced than a chimpanzee's for the first three to four million years of hominin evolution. *Homo habilis* persisted until 1.44 million years ago (Spoor et al., 2007).

Oldowan tools were made by taking one stone and using it to knock off flakes from another stone. The sharp-edged flakes as well as the stones from which they were chipped off were used as cutting tools. Scrape marks on animal bones show that they were used to cut meat off bones. Discarded Oldowan stone tools were found with these bones and with bones that were split open to extract marrow (Toth and Schick, 1993). Over more than a million years, one Oldowan tool looks much like another—nothing changed.

125

Very, Very Slow Change

The first recorded sign of a creative impulse was apparent 1.7 million years ago with the Acheulian technology, associated with *Homo erectus*. *Homo erectus* had bigger brains and body proportions closer to ours. Figure 5-2 shows the evolution of stone tools from the Oldowan chopping tools (A) to symmetrical Acheulian lower Paleolithic "hand axes" and scrapers (F). The technology involved chipping away, one flake at a time, from a core. It remains the method by which most sculptors work. (Some contemporary sculptors even use chain saws on wood.) It requires advance planning and a clear idea of the finished product. As a tool-maker chipped away, he also would have to take into account the effect of the previous chip of stone before flaking away the next bit of stone. If we were writing a "grammar," a set of instructions for making Acheulian tools, it would

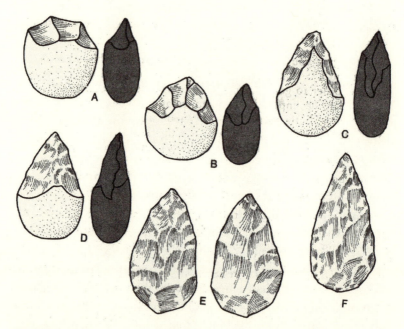

Figure 5.2. Evolution of stone tool technology from Oldowan chopping tools to Acheulian "hand axes." Reprinted from Lieberman, 1975. Originally redrawn from *The Old Stone Age*, by F. Bordes © 1968.

have to be "context sensitive," taking into account the result of the previous steps: each time that the toolmaker picked up his hammer-stone and knocked off another stone flake. This isn't conceptually different from the grammars that linguists devise when they attempt to describe how sentences are formed.

It's not clear whether so-called hand-axes were ever used to chop anything. More likely, they were used to cut and scrape. Acheulian technology became entrenched and was the dominant method of stone-tool production for the next million years. Chipping away isn't as efficient as the later Levalloisian technique, which can mass-produce finished tools while conserving raw materials. Long, narrow-bladed knives and scrapers can be rapidly produced. Levalloisian stone tool-making first entails chipping away at a large stone to pre-pare a core that doesn't resemble any tool at all. The core perhaps best resembles a headless, footless stone turtle. The next step requires a conceptual set-shift. When you have decided that your stone turtle-like core looks right, you have to switch to a different technique. You can either apply pressure to the core with a stick or strike it briskly with a hammer-stone in such a way that a finished tool fractures from the core on a fault line. The Levalloisian tool-making technique thus entails knowing that stones will cleave on fault lines. This knowledge base and technique must be transmitted through the medium of lan-guage. Whereas an adult chimpanzee can sit beside a young chim-panzee and tutor it to crack nuts successfully without uttering a word or using sign language (Boesch, 1993), it's difficult to see how you could convince a chimpanzee to make a turtle-like core without ex-plaining what it could be used for.

The Levalloisian technique clearly entails a conceptual "jump" to produce the finished product. The jump is arguably greater than those typical for the generative grammars constructed by Noam Chomsky and his adherents, where you are simply rewriting one sentence into another. It is similar to making a cake. The initial steps in cake making involve mixing and stirring flour, milk, eggs, sugar, and so on, into a thick batter that has no resemblance whatsoever to a cake. After the batter is formed, it's put into the oven and baked. If the time and oven temperature are right, you have a cake.

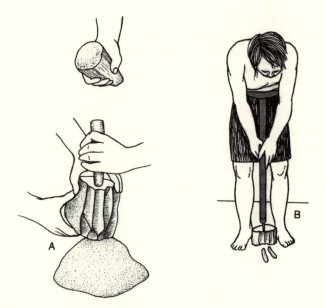

Figure 5.3. Levalloisian technique (A) simple and (B) advanced. Levallosian technology involves a conceptual technique that permits mass production. Reprinted from Lieberman, 1975. Originally redrawn from *The Old Stone Age*, by F. Bordes © 1968.

Figure 5.3 depicts the end-stage of the Levalloisian technique. The Levalloisian technique was introduced in the Middle Paleolithic about 200,000 years ago. However, the older Oldowan technique continued to be used. It still was in use in 1832 when Charles Darwin arrived in Tierra del Fuego on the *Beagle*. However, the indigenous people living in Tierra del Fuego observed by Darwin in 1832 had no difficulty learning English, nor do their immediate descendants have difficult using contemporary tools. That the stone-tool technology at Tierra del Fuego was at the Oldowan stage didn't signify that anyone there had *Homo habilis* brains.

Moreover, the Levalloisian technique continued in use as an important industrial process up to the early years of the nineteenth century. During the Napoleonic wars, millions of flints for muskets were mass-produced by artisans striking rocks with little hammers. The "simple" Levalloisian technique is in use today to cut the shiny facets on gemstones.

Therapsid Bones, Human Tongues, and Brains

Anatomy as a Window on the Brain

Anatomy can provide the key that unlocks the nature of the brains of long-dead creatures. As chapter 1 noted, the anterior cingulate cortex (ACC), which becomes active when you direct attention to virtually anything, whether it is Jesus's miracles or your tax return, dates back to Therapsids, mammal-like reptiles who lived in the age when dinosaurs roamed the earth. The soft tissue of the brain obviously does not survive 260 million years. The inference that Therapsids had an ACC is based on their fossil remains having the three middle ear bones found in all present-day mammals. As you may recall, in reptiles these three bones form a jaw-hinging mechanism that allows snakes to open their jaws wide to swallow large creatures, as the snake-fancier's python was doing as it was swallowing his arm in his knotty-pine paneled sunroom. Anatomy provides the key to inferring that Therapsid brains had an ACC. Anatomy—the human tongue—allows us to trace back the time at which we can be certain that hominins had brains more or less like yours and mine.

The Human Tongue

The peculiar anatomy of the human tongue, which reflects "re-design" for talking, can tell the story of when we had brains fully capable of talking and being unpredictably creative. You may not be aware that your tongue is very peculiar because hardly anyone examines the tongues of cats, dogs, cows, or apes. Victor Negus's studies, however, showed that at birth your tongue was similar to an ape's, located almost entirely in your mouth (Negus, 1949). But it gradually changed, moving down into your throat and changing its shape until sometime between age six and eight years it was unlike that of any other creature on earth (D. Lieberman et al., 2001).

Charles Darwin first raised the question of why we have peculiar tongues. In the first edition of *On the Origin of Species*, he noted

> the strange fact that every particle of food and drink which we swallow has to pass over the orifice of the trachea, with some risk of falling into the lungs (Darwin, 1859, p. 191).

Humans are susceptible to choking on food because when we swallow we have to execute strange acrobatic maneuvers to prevent food from falling into our larynx. That's because as the human tongue moves down into the throat between birth and age six to eight years, it carries the larynx down with it. In contrast, in human newborn infants and most other creatures throughout life, the larynx is close to the opening to the nose and can lock into the nose when breathing, yielding a sealed tube within a tube for breathing.

Figure 5.4a shows the tongue, mouth, and larynx of a human newborn. At birth, the newborn infant tongue is positioned almost entirely in the mouth; the larynx is close to the opening into the nose. This arrangement allows a newborn infant to simultaneously suckle milk and breathe. The infant's larynx rises and locks into the nasal passageway like a small periscope, sealed off from the pathway in which water, milk, and soft solids move past on either side of the raised larynx. Apes, dogs, cats, and other mammals have similar tongues and mouths throughout life. That's why dogs and cats can slurp away without stopping to breathe until their bowls are emptied. Newborn humans are obligate nose breathers until about age three or four months, when tongue, throat, and skull anatomy begin to change. Anatomy and neural control are matched to facilitate an infant's being able to feed without risking choking.

Figure 5.4b shows an adult human tongue, airway, and the position of the larynx and entrance to the esophagus. The esophagus leads to the stomach. When it comes to swallowing solid food, apes and most other animals also have an advantage. Their tongues first propel food along the roof of the mouth, past the larynx, and into the esophageal pathway leading to the stomach. As Daniel Lieberman points out in his 2011 book *The Evolution of the Human Head* (pp. 295–302), swallowing and breathing patterns differ profoundly between humans and nonhuman primates. In nonhuman primates, the pharynx forms a sort of "tube within a tube" in which air flows directly from the

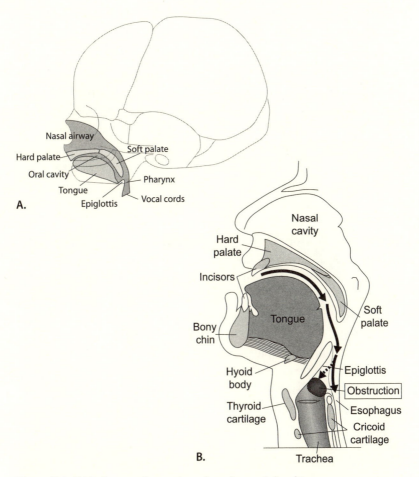

Figure 5.4. (A) Infant vocal tract. A newborn human's head in many ways resembles a hairless infant chimpanzee's. That's also the case for the infant's tongue and the position of its larynx (the voice box) and vocal cords. The newborn human tongue is positioned almost entirely within the mouth, and its larynx, which connects to the root of the tongue, is close to the opening to the nose. (B) Adult human tongue and airway. The human tongue after age six to eight years is different from that of any other living species. Half of the human tongue is positioned in the mouth, half is in the pharynx, down in the throat. The airway above the larynx formed by the tongue positioned in this manner has two equal-length segments that meet at a right angle. This configuration permits us to produce vowels that enhance the intelligibility of human speech, but there is a biological cost. The human larynx has been carried down into the throat because it is connected to the root of the tongue. The low position of the adult human larynx increases the risk of choking on food and causes swallowing disorders.

131

lungs through the nose. Food can't interfere with the air flow through this inner pathway. The larynx, the entry to the inner air pathway, is positioned high, within the opening to the nose, sealed off from the outer tube through which liquids and soft solids are swallowed. Two soft tissue"flaps," the epiglottis and velum, overlap each other (think of the rubber gaskets on a refrigerator's door) to seal off the raised larynx from the food pathway when drinking. When swallowing solid food, the epiglottis flips down, and breathing stops momentarily. The increased risk of choking faced by older humans derives from the human tongue having gradually changed its shape, migrating down into the pharynx in the first six to eight years of life, carrying the larynx down with it.

When swallowing, we have to perform a set of coordinated maneuvers to avoid choking. We have to pull the larynx and the hyoid bone (which supports the larynx) forward and upward to get it out of the way of solid bits of food that are being forcefully propelled down our pharyngeal air-food pathway. At the same time, we have to flip our epiglottis down to cover the larynx. About 500,000 people in the United States have trouble performing this circus act and suffer from swallowing disorders (*dysphagia*). In most restaurants, you can find a poster in the kitchen showing the staff how to perform the Heimlich maneuver, which attempts to clear the larynx by "popping-out" food blocking a diner's larynx. Applying pressure to the choking person's abdomen compresses the lungs, which hopefully will push the food out of the larynx. However, it doesn't always work, and asphyxiation from food lodging in the larynx still remains the fourth largest cause of accidental death in the United States (National Safety Council, 2010).

Growing a Human Skull, Neck, and Tongue

It takes six to eight years, sometimes ten, for the human skull, tongue, and neck to approximate their adult proportions and shape. Daniel Lieberman and his colleagues tracked the descent and reshaping of the human tongue using cephalometric radiographs (the x-rays that orthodontists use when children have braces) of 15 males and 13 females that were taken at three- to six-month intervals between the ages of one month and 14 years. The x-ray series was made in the 1940s and 1950s, before the adverse effects of radiation were apparent. Digital instrumentation enabled them to make precise measurements (D. Lieberman and McCarthy, 1999; D. Lieberman et al., 2001). The oral cavity's length shortens in the first two years after birth through differential bone growth that moves the hard palate (the roof of the mouth) backward. The tongue begins to move down into the neck, carrying the hyoid bone and the larynx down with it (the larynx is suspended down from the hyoid). As the tongue descends, the length of the neck gradually increases.

This developmental process yields the species-specific human supralaryngeal vocal tract (SVT), the airway above the larynx. As chapter 2 pointed out, vowels and consonants are specified by their formant frequencies—the frequencies at which maximum acoustic energy can pass through the supralaryngeal vocal tract. The shape of the SVT determines the formant frequencies in much the same manner as the shape and length of an organ pipe determines the note that it produces. When a child's SVT has reached its adult proportions, half of the tongue is positioned in the mouth, half in the pharynx. The pharynx is the inside of the neck. The posterior, back contour of the human tongue has become circular, and the oral "horizontal" SVTh and pharyngeal "vertical" SVTv segments of the tongue meet at a right angle

Why Apes Can't Talk

Chance events can set the course of your life. We were living in Lexington, Massachusetts, and our home, which was a former carriage house, had an old fashioned, long bathtub. I was stretched out in the warm water, languidly listening to Bill Cavness's weekly broadcast "Reading Aloud" on the Boston NPR station WGBH. The book that Cavness was reading was a collection of essays by the distinguished anthropologist Loren Eiseley. In passing, there was a sentence about apes not being able to talk. I wondered, why? I had been working on respiratory control in speech and other aspects of speech, but I hadn't really thought about why we, and no other creature, talks.

My first attempt to answer why apes can't talk was to find out what apes actually produced when they vocalized. We moved the next year to Storrs, Connecticut, where Marcia and I both taught at the University of Connecticut—she in the Department of English and I in the newly formed Department of Linguistics. There were close ties between research at UConn and the Haskins Laboratories, then located in New York City. The Haskins Laboratories had been founded by the philanthropist Caryl Haskins, who was an eminent scientist in his own right, shortly after the end of World War II. I was supposed to be studying speech motor control in humans. Nobody else at Haskins seemed to be particularly interested in why apes could not talk, but neither did the Laboratories' research directors, Alvin Liberman and Franklin S. Cooper, object to my looking into the matter. So I bought a portable, briefcase-sized reel-to-reel Sony tape recorder at one of the nearby photo and electronics stores and arranged to record chimpanzees at the Prospect Park Zoo in Brooklyn. I still have the Sony tape-recorder; perhaps it will end up in a museum.

It was winter, and the apes and other animals were in the same building. I recorded lots of ape calls. The zoo staff was very helpful, making sure that I recorded all the calls that they had heard the chimpanzees produce. At Haskins, I then analyzed the set of calls using the Sound Spectrograph, at the time the primary research instrument for acoustic analysis. The Sound Spectrograph uses a whirling metal drum to record a two-second segment of speech and a filter that, driven by a gear on a long vertical axis, shifts through the range of frequencies being analyzed. The Sound Spectrograph had been designed in the 1930s at the Bell Telephone Laboratories and was still in use until computers began to replace it for speech analysis in the 1970s.

The analysis procedure was slow. The spectrograph's output, on spectrograms, blackened areas on pieces of paper, was produced by a technique that dated back to the 1920s to transmit "wirephotos."

The analysis took several months, and would not have succeeded without the advice of Franklin S. Cooper, who had viewed thousands of spectrograms. We were at first puzzled by some strange spectrograms until we realized that Mynah birds who were in the same zoo house had been imitating the chimpanzee calls. The resulting study, which appeared in 1968 in the *Journal of the Acoustical Society of America*, showed that all the ape calls seemed to be produced with their tongues producing a vocal tract shape that resembled a slightly flared tube, sometimes constricted by their lips. In short, the chimpanzees were producing grunts that didn't deviate from the human "uh" sound, except for consonantal sounds resulting from constricting their lips that were sort of "w"-like. The pitch of the chimpanzee cries, however, went up and down. They also produced hoarse screams.

But the possibility remained that zoo apes don't use their full anatomical possibilities. So we (by this time Dennis Klatt, a speech researcher in Kenneth Stevens's MIT laboratory, was involved) used then state-of-the-art computer-modeling techniques to explore the full range of anatomical possibilities that monkeys, apes, and newborn human infants could employ to modify the shape of their supralaryngeal vocal tracts. Other researchers joined in, including Katherine Harris at Haskins; Bill Wilson, a primatologist; and Edmund S. Crelin, the Yale University anatomist who in 1969 had published the first-ever comprehensive anatomical study of human newborn infants. We could have pounded brass tubes into the range of shapes that nonhuman SVTs could assume and then determined their acoustic filtering properties. The tubes would be similar to the vox humana pipes on some pipe organs, but the computer technique allowed us to systematically model the full range of possibilities by plugging in numbers that produced the range of shapes that were anatomically possible.

Our computer-modeling studies showed that unless you have an adult-like human tongue and vocal tract, you inherently cannot produce the full range of human speech (Lieberman et al., 1969, 1972; reviewed in Lieberman, 2012). It is impossible to produce the vowels [i], [u] ,and [a] (the vowels of the words "see," "do," and "ma"). These vowels are among the few attested "universals" of human language, delimiting the range of vowels used in all human languages (Greenberg, 1963). At MIT, Ken Stevens showed that [i] and [u] had the lowest chance of being confused with another vowel. At the Bell Telephone Laboratories, Peterson and Barney had come to the same conclusion. The initial focus of the Bell Telephone project was to develop a system that would enable voice-controlled telephone "dialing." The system obviously would have to work when anyone attempted to place a call. At the Bell Telephone Laboratories, panels of listeners therefore were placed in the position of an automatic system that wouldn't be familiar with the voice of the person attempting to phone, or able to see him or her. Under these conditions, the vowel [i] was confused two times in 10,000 trials, [u] six times. Other vowels were confused hundreds of times (Peterson and Barney, 1950). Almost 50 years later, Hillenbrand et al. (1995) replicated Peterson and Barney's findings, using computerized analysis techniques. The speakers studied by Hillenbrand and his colleagues even spoke the same dialect of English, but that didn't help. Words that differed with respect to vowels other than [i] and [u] were misidentified hundreds of times.

Our initial 1969 monkey study used a computer-implemented vocal tract modeling technique that calculated the formant frequency patterns of the vowels that a rhesus macaque could produce. The range of tongue shapes that a monkey could produce was estimated by taking into account constraints on tongue deformation (since confirmed by Takemoto, 2001, 2008) and moving its tongue toward the positions used by adult human speakers to yield the vowels [i], [u], and [a]. The computer model showed that the monkey's SVT precluded producing these vowels. Subsequent computer modeling studies revealed similar phonetic limits for chimpanzees and human newborn infants. Newborn humans have SVTs similar to those of nonhuman primates (Negus, 1949; Crelin, 1969). Our modeling studies took into account the dynamic motions of the tongue, lips, and larynx that were evident in the cineradiographic Truby et al. (1965) study of newborn infant cry.

Quantal Vowels Enhance Human Speech

Kenneth Stevens at MIT in 1972 provided part of the answer to why the vowels [i], [u], and [a] make human speech a more effective means of vocal communication. Stevens showed that the species-specific human SVT can produce the ten-to-one midpoint area function discontinuities that are necessary to produce these vowels, which Stevens termed "quantal." In the species-specific human vocal tract, the back contour of the tongue is almost circular. Half of the tongue, (SVTv) is positioned in the pharynx; half of the tongue (SVTh) is positioned in the oral cavity. SVTv and SVTh meet at an approximate right angle, owing to the tongue's posterior circular shape. The extrinsic muscles of the tongue, muscles firmly anchored in bone, thus can move the undeformed tongue to form abrupt midpoint ten-to-one discontinuities in the cross-sectional area of the SVT. Stevens, using both computer-modeling and wooden models of the SVT (having the shapes of strange woodwind instruments) showed that these abrupt midpoint discontinuities were necessary to produce [i], [u], and [a]. The quantal vowels are perceptually salient because two formant frequencies yield spectral peaks that make it easier to differentiate spoken words. A visual analogy would be allowing people to communicate using signal flags that had bright, saturated colors in place of flags that had pale colors The formant frequency patterns of quantal vowels also do not shift when tongue position varies slightly about the midpoint—speakers can be sloppy and produce the "same" vowel.

In 1978, Terrance Nearey showed that the vowel [i] and to a lesser degree the vowel [u] had another property that explained why [i] and [u] were almost never confused with another vowel. The vowel [i] is an optimal signal for determining the length of a speaker's vocal tract—a necessary step in the complex process of recovering the linguistic content from the acoustic signals that convey speech (Nearey, 1978). Human listeners, as well as the computer systems used in telephone systems that attempt to recognize what you're saying on a telephone, have to guess at the length of your SVT. This is necessary because the formant frequencies for a speaker having a very short SVT will differ from those produced by a speaker having a long SVT for the same

137

word. We don't notice this. We "hear" the same words when a person who has a long SVT or short SVT is talking, even though the absolute values of the formant frequencies are very different, because our brains automatically and unconsciously compensate for the different SVT lengths. Nearey showed that the [i] is the optimal vowel for doing this.

Independent computer-modeling studies (Lieberman et al., 1972; Lieberman and Crelin, 1971; Stevens, 1972; Carre, Lindblom, and MacNeilage, 1995; Beckman et al., 1995) have replicated these results. The most recent independent study, De Boer (2010), again shows that the "quantal" vowels [i], [u], and [a], cannot be produced unless SVTv and SVTh have equal lengths and meet at a right angle. That entails having an adult-like human tongue. But these studies all suggest that you also needed a brain that could control the SVT to produce human speech. All nonhuman primates and many other species have tongues that could produce enough vowels and consonants to enable them to talk—if they only had a brain that could learn and execute the complex motor control maneuvers that are necessary to talk. They could not produce quantal vowels and some quantal consonants (cf. Stevens, 1972; Lieberman, 1984, 2000, 2006, 2012), but they can produce enough vowels and consonants to talk, albeit less clearly. This includes Neanderthal and earlier hominins. We don't have recordings of Neanderthal's talking, but as Edmund Crelin and I pointed out in 1971, the archaeological evidence shows that they must have talked and possessed language. Our human capacity to produce quantal vowels yields an incremental selective advantage for human speech. However, it is evident that apes don't talk—therefore they don't have brains that can either learn or execute the rapid articulatory maneuvers that are involved in talking. This, of course, fits into the FOXP2 story, but that was about 40 years into the future.

Neanderthals

In the fall of 1970, I walked into Edmund S. Crelin's office at Yale University's Department of Anatomy. Ed had recently published the first comprehensive anatomy of the human newborn, and mutual friends had suggested that I see him. I was carrying plaster copies of the fossil skull and jawbone of the Neanderthal man unearthed near the French village of La Chapelle aux Saints at the start of the twentieth century. Ed's reaction when he looked at the skull was, "It's a big baby." Muscles leave marks on bone, allowing the soft tissue of the Neanderthal to be reconstructed. The almost total similarity of the base of the human newborn skull and the Neanderthal fossil guided Crelin's reconstruction, which in essence ended up looking like a very, very, large human newborn infant SVT.

Our computer modeling study of the reconstructed Neanderthal SVT showed that though it clearly could have produced a range of speech sounds that would have enabled him (it was an adult male's skull) to talk, the quantal vowels [i], [u], and [a] would have been absent. This did not mean that Neanderthals did not possess speech or language. Based on the available archaeological evidence, we concluded that Neanderthals possessed language and could talk, albeit not with the same clarity as modern humans. Our 1971 paper "On the Speech of Neanderthal Man" started a firestorm. Defenders of Neanderthals, who for the most part didn't seem to have actually read our paper, declared that we claimed that Neanderthals were brutes who did not have language. A subsequent stream of scholarly papers has both denounced and replicated our conclusions. The most recent replication is De Boer's 2010 independent modeling study, which shows that a vocal tract that doesn't have equal length sections of the tongue in the throat and mouth (SVTv and SVTh) cannot produce quantal vowels.

I must also stress the fact that [i] and [u] are not the only speech sounds that allow us to unconsciously estimate the length of SVT of the speaker we're listening to. It's possible to estimate a speaker's SVT length by listening to a short length of speech and taking account of the average values of the formant frequencies (Ladefoged and Broadbent, 1957). Other speech sounds are also useful, and if you already know what someone is probably saying (such as the conversational openers "hi," "hello," "good morning," and so on), you also can "reverse engineer" the speech recognition process and guess at the speaker's height because vocal tract length correlates with height.

Humans and dogs (Kaminski et al., 2004) — and probably other animals — are able to recognize words because in the course of evolution they developed neural mechanisms that used formant frequencies of vocalizations to estimate the size of whoever was vocalizing. The process works best if the vowel is always fixed, and that's the case for all of the animal vocalizations analyzed by W. Tecumseh Fitch and his colleagues. Contrary to what Fitch often states in his lectures and in his 2010 book, the SVTs of the animals he has studied are incapable of producing quantal vowels because they lack 1:1 SVTh/SVTv tongue segments that meet at a right angle. All of the animal vocalizations shown by Fitch in his 2010 book *The Evolution of Language* are "uhs" — the "schwa" vowel that people utter when they're tired or trying to think of the next word. Fitch in his 2010 article with Brant de Boer aimed at a technically competent audience curiously agrees with the view that nonhuman tongues and SVTs cannot produce the quantal vowels [i], [u], and [a].

In short, speech communication wouldn't be as effective, but it would be possible, absent quantal vowels. That was the conclusion of the 1971 Lieberman and Crelin paper "On the Speech of Neanderthal Man." We explicitly stated that Neanderthals had speech and language, but their speech was not fully human; Stevens, Nearey, and other researchers then gradually showed why the human SVT enhanced the robustness of speech communication. But Neanderthals continue to be a puzzle that we will return to shortly.

Speech Is Special

Many proposals for the evolution of human language assume that hominins first used manual gestures to communicate rather than talking. As I pointed out earlier, apes can be taught to use manual sign language or use keyboards to communicate using words, but they cannot talk. Given the complexity of speech, why do we instead talk? The answer rests in some obvious factors, and one that became apparent during the course of what was supposed to be a straightforward engineering project. Talking clearly frees your hands so you can use tools, throw things, or perform the many tasks that only hands can do. You can also communicate in darkness, and you don't have to keep looking at whoever is signaling.

However, there is another advantage to talking. Human speech allows us to rapidly transmit information. Speed is a critical component of all information transfer systems—that's why billions of dollars are spent each year continually upgrading Internet transmission systems, switching to faster and faster fiber-optic systems. A group at Haskins Laboratories was developing a machine in the 1960s that would "read" printed texts aloud to blind people. One of the Laboratories' prime objectives was to develop a text-to-speech reading machine. Caryl Haskins, who had founded the Laboratories, had thought that goal could be attained using then existing technology. He was correct insofar as it wasn't difficult to convert orthography to a phonetic code, but it soon became clear that it wasn't a simple matter to convert the phonetic code to speech.

Linguists since the time of Sanskrit scholars thousands of years ago had thought that when we talked we simply strung together phonemes, elemental sounds roughly approximated by the letters of the alphabet, as though they were beads on a string, a sort of movable type. The phonemes [t], [a], and [b], for example, can be rearranged to form "tab," "bat," "at," "ba." So it should have been possible to build a mechanical synthesizer that would talk, once you had isolated the phonemes of English. However, much to everyone's surprise, it proved to be impossible to isolate a set of sounds that corresponded to letters that could be permuted to form words. Tape recording, the latest form of sound recording in the 1960s, seemed to offer a simple solution. Everyone thought, following the advice of linguists, that it would be possible to isolate phonemes, equivalent to the letters of the alphabet, from tape recordings. When a person spoke the word "too," there surely must be a segment of tape that contained the sound [t], before a segment of tape that contained the sound [u] (the phonetic symbol used by linguists that is the sound of the letters "oo"). Attempts were made to isolate clearly enunciated sounds by cutting out bits of tape from recordings made by trained radio announcers. However, when the

141

segment of recording tape that they thought corresponded to the "phoneme" [t] in the word "too" was isolated and linked to the vowel [i] segmented from the word "tea," the resulting acoustic signal was incomprehensible. One leading researcher declared that a test sentence constructed from ostensible phoneme-segments cut out of tape recordings sounded like "a drunken cockroach talking." (He never explained when pressed what a cockroach, drunken or not, would sound like.) It became evident that there are no "pure" speech sounds because the formant frequency patterns of the hypothetical independent phonemes were melded together into syllables and words.

For example, the formant frequency patterns that convey the phonemes [t], [ae], and [p] of the word "tap" are melded together into one syllable. As the tongue moves from its position against the roof of the mouth (the palate) to produce the syllable-initial "stop" consonant [t], a formant frequency pattern is produced that transitions into that of the [ae] vowel, and then to the final stop consonant [p]. Human speakers plan ahead. If instead you say the word "too" (in phonetic notation [tu]), your lips are already "rounded" (protruding and narrowing), anticipating the [u] vowel. Your lips are not rounded at the start of the word "tea," because the following vowel [i] does not require lip rounding. Look at your lips in a mirror when you say "tea" and "to" (in phonetic notation [ti] and [tu]). Your lips will be rounded when you say "too," but not when you say "tea." This inherent "coarticulation" of speech is its inherent value in the struggle for existence. As we talk, we circumvent the data rate imposed by our ordinary, garden-variety, mammalian auditory system.

The Haskins speech research group directed by Alvin Liberman found that when people listen to a stream of individual nonspeech sounds, they fuse into a buzz when the rate exceeds 15 sounds per second. At slower rates, it still is almost impossible to make out what the sounds are. Transcribing messages coded by alternative sounds such as Morse code is slow and requires all your attention. In contrast, speech allows humans to transmit phonetic distinctions at rates of up to 20 to 30 "segments" per second. Talking achieves this rapid transmission rate because human speech is an "encoded" signal in which information is transmitted at the slower syllable rate, then "decoded" into phonetic segments by means of perceptual processes in our brains whose nature is still being debated (Liberman et al., 1967). If you were limited to the slow nonspeech rate and were listening to this sentence, you would forget the beginning of this sentence before you came to its end. Speech thus plays a critical role in the evolution of human language, making complex syntax productive.

A linguist at a conference once thought that he was insulting me when he declared that "Lieberman is only an engineer." My first two degrees from MIT are in electrical engineering, and I was pleased, because engineering solutions have to work. Engineering projects directed at automatic speech recognition came to the same conclusion as the Haskins research group. When you're talking to an automated telephone-answering system or your smart-phone, it is matching your incoming signal to complete words stored in its memory bank, not phonemes. Linguists have not caught up with engineers and don't realize that their smart-phones and computers are recognizing complete words, rather than attempting to segment the speech signal into phonemes, then grouping the phonemes into words. The computer systems, like human listeners, must take into account the effects of different supralaryngeal vocal tract lengths, which shift the formant frequency patterns of words, as well as dialect differences.

Encoding differs somewhat from language to language. For example, Swedish speakers plan ahead for a longer interval at the start of a syllable than English speakers (Lubker and Gay, 1982). The minimal speech encoding unit appears to be the syllable, but encoding effects can have longer spans. We are not aware of speech encoding because our brains perceptually recover segmental phonemes. English also uses alphabetic orthography, so it's "natural" to assume that somehow there are segments of the acoustic speech signal that correspond to the letters of the alphabet. However, orthographic systems that code entire words, such as is the case for the Chinese languages, are used by more than a billion people.

When Did We Have Fully Modern Brains?

The significance of the probable speech limitations of Neanderthals paradoxically rests in what it suggests about the neural capacities of our immediate ancestors, rather than Neanderthal speech per se. Robert McCarthy and his colleagues (including me) used a different approach than the 1971 Lieberman and Crelin study to determine when fully human tongues evolved. Rather than attempting to establish the shape of the tongue and SVT from skeletal features similar to those of newborns, we followed up on the suggestion made in my 1984 book (pp. 292–296) on the limits placed on tongue proportions by the length of a person's neck. The details are in McCarthy et al., (in preparation) and Lieberman (2012).

Sandro Botticelli's Long-Necked Beauties

Long necks turn out to have a purpose, other than beauty—perhaps we view long necks as a sign of beauty because they make it possible for us to produce clear human speech that has quantal vowels, while enabling us to swallow. In brief, the procedure used by Rob McCarthy and colleagues involves first determining the length of the horizontal component of the tongue (SVTh), the segment that is in the mouth. SVTh can be estimated from a fossil skull using highly visible bony landmarks and adding 10 millimeters (mm) to approximate the lips. The length of SVTv, the portion of the tongue in the pharynx, is constrained by the fact that everyone has to be able to eat. In human children, as the tongue moves down into the pharynx between birth and age six to eight years, it carries the larynx down with it into the throat. As this happens, children's necks gradually lengthen to accommodate the tongue and larynx. A study published 23 years after the Lieberman and Crelin Neanderthal paper established neck lengthening in children (Mahajan et. al., 1994). The point of neck lengthening is to allow us to swallow even though our tongue is carrying the larynx down toward the chest during the first six to eight years of life. The cricoid cartilage of the larynx (the Adam's apple) in a human adult reaches a position opposite the sixth or seventh cervical vertebra of the neck (Hiiemae et al., 2002). If it were to fall much below this point, it would be impossible to swallow because the larynx would be positioned in the chest, blocked by the collarbone when the hypothetical hominin attempted to move it forward and upward to avoid having food lodge in its larynx.

In other words, your neck has to be long enough to accommodate a tongue that dragged the larynx down into it. Robert McCarthy measured the cervical vertebrae of the necks of 73 specimens of modern humans from populations distributed around the globe, as well as a sample of Neanderthals (La Ferrassie, Le Moustier, Shanidar 2) and Upper Paleolithic fossils such as Cro-Magnon, who are considered to be virtually similar to modern humans, and the earlier Skhul V fossil, who lived about 90,000 years ago. The names used to identify these fossils are usually the place-names of the locations where they were unearthed. The Cro-Magnon fossil is only one early specimen of an Upper Paleolithic fully modern human. Skhul V is a Middle Paleolithic fossil much closer to modern humans than Neanderthals. Neck lengths were then estimated from the dimensions of the recovered fossil vertebrae, "tilting" the estimates toward modern human neck lengths by using modern human discs—the discs that separate vertebrae. In cases of doubt where individual vertebrae were missing from a fossil, maximum-sized vertebrae from modern humans were used.

The "hard" fossil evidence that precludes Neanderthals having the 1:1 SVTh/SVTv ratio necessary to produce quantal vowels is the long Neanderthal mouth. The distance between the bony landmarks on the base of the skull (prostion and the basion), which sets the limit on SVTh, is outside the range noted by Howells (1989) for 2,504 adult human skulls drawn from ethnic groups throughout the world. When the Neanderthal SVT was modeled, its SVTh/SVTV proportions absolutely prevented it from generating quantal vowel formant frequencies, as Stevens (1972), Carre et al. (1995), Beckman et al. (1995), and De Boer (2010) independently confirm.

Surprisingly, a short neck length also would have prevented Skhul V, a Middle Paleolithic fossil that is generally believed to be closer to the line of human descent, from producing fully quantal vowels.

With uncertainties skewed toward values found in modern humans, neck length averaged 120.3 mm in Neanderthals (very close to 115 mm estimated for the Neanderthal Shanidar 2 fossil by Trinkhaus in his 1983 study), 109.1 mm in Skhul V, and 133.7 mm and 127.0 mm in two modern human samples. When the larynx is placed in the neck position (opposite the sixth or seventh cervical vertebra) that it has in a human adult, the Neanderthal vocal tract has a 1.5 SVTh/SVTv ratio similar to that of newborn to two-month-old human infants. Placing the Neanderthal larynx at the lowest point that would permit swallowing resulted in a 1.3 SVTh/SVTv ratio. That ratio is in the range of two-year old children (D. Lieberman et al., 2001; Lieberman and McCarthy, 2007) and cannot produce fully quantal vowels. In part, two-year-olds sound like two-year-olds because they cannot produce quantal vowels. Patricia Kuhl and her colleagues (1992) showed that we "hear" young children producing [i]s through a perceptual process that convinces us that they are, when they actually aren't. But the children's [i] approximations are neither as stabile or furnish as useful information on SVT length as true [i]s. Some phoneticians may demur—claiming that their ears are as sensitive as computer-implemented speech analysis, but the comparative Kuhl et al. (1992) study shows that human speech perception is influenced by age six months by the language or languages that we have heard.

In short, the vocal tract anatomy necessary to produce quantal vowels becomes evident in Upper Paleolithic fossil hominins who lived about 40,000 years ago. Since natural selection can act only on overt behavior, we can be certain that they possessed neural circuits that enabled them to use their peculiar human tongues to full advantage; else there would have been only negative consequences (choking to death) to having human SVTs. Thus, through the lens of anatomy, we can discern the presence of the brain mechanisms that regulate the voluntary, rapid, complex, internally guided motor acts that underlie human speech. Brains and body coevolved to make human speech possible. But we can infer more than that, since the cortical-basal ganglia circuits that allow humans to learn and sequence the complex motor acts involved in speech also act in the realm of cognition. The burst of technological change and the appearance of art in the European Upper Paleolithic has often been interpreted as evidence for a "cultural revolution" (Klein, 1999), signifying an abrupt increase in hominin cognitive capabilities at that time. However, that cannot be the case.

Human SVTs and the neural bases for speech and language must have been present in Africa well before 40,000 years ago—probably 260,000 years into the past. Every human being on earth has African ancestors, though some left Africa earlier than others. The ancestors of the "non-African" (European) humans whose fossil remains Rob McCarthy examined had emigrated from Africa to Europe, displacing the indigenous Neanderthal population. As present-day colonists do, they brought their own culture and toolkit with them. Moreover, not everyone left Africa during the successive migrations that took place when Europe and Asia were first colonized.

It's clear that contemporary Africans have perfectly human SVTs (McCarthy measured neck lengths in different groups of human populations, including Africans), and present-day African languages use quantal vowels. And any child of recent African descent (we are all ultimately Africans) can master any language on earth, and has normal cognitive capacities, barring insult to the brain that can strike anyone.

Fashion—An Infallible Index of Creativity

The established procedure for studying the evolution of the bones, muscle, and behavior of any species can provide guidance as to what constitutes evidence for behavior that reflects creativity. Simply put, we can start by identifying aspects of human behavior, absent in our nearest primate relatives, chimpanzees, that meet two criteria: (1) everyone agrees that they reflect the creative impulse that can manifest itself present in human society, and (2) they must leave traces visible over vast stretches of time.

Chimpanzees make and use simple tools (Goodall, 1986), so tools, though suggestive, are not an infallible index of mind/brains capable of the creative acts that mark humans and make us so unpredictable. Chimpanzees can also learn human sign language to a limited degree through observation and imitation (Gardner and Gardner, 1969) and use a few vocal calls to signal what they want (Slocombe et al., 2010). So even if we could somehow find irrefutable evidence for early hominins using simple forms of language, that would not show that they had fully modern human brains.

We have to look for behavior that is absent in chimpanzees. One element absent in chimpanzees is our preoccupation with the range of things and activities that constitute "art"—music, dance, recitals, paintings, performances, and yes, fashion. Fashionable objects boil down to art that is accessible to most everyone. At the annual street fair of the Rhode Island School of Design, one hundred or so artists have their work on display for sale on four streets blocked off to traffic. There are a few stalls showing paintings—of flowers, dogs, trees—but "high-art" isn't on display. For the most part there are hats, scarves, and other items of clothing and jewelry. Everything is always in fashion.

Here, you the reader well may ask, what differentiates art from fashion? I can't tell you, other than to state that people are drawn to useless objects that some consider fashionable gewgaws, others, works of art. There are beautiful hand-blown glass cups for sale, but a beautiful cup doesn't facilitate being able to drink water or anything else. You might wonder if hats, scarves, jeans, or woven jackets constitute an art form, but where can the distinction be drawn? What makes one poster an advertisement, a poster by Marc Chagall a work of art?

Price isn't the answer, because the art market is fickle. Victorian tapestried sofas that once sold for the equivalent of hundreds of thousands of dollars are now almost worthless curiosities. Benjamin Disraeli, the Prime Minister of the United Kingdom in the late Victorian age, was a few years ahead of the pack when he directed the British Museum to purchase the Greek art of the "golden age," which then was not valued. Disraeli was almost voted out of office for squandering the nation's treasure before the museum's purchases became "priceless, timeless" art. What "objective" criterion defines a weaving displayed in a museum as a work of art and a pair of pants as a pair of pants? If the pants were selected by a museum curator, would they be a work of art? The boundary between fashion and art is in the eye of the beholder. I see fashion as a reflex of our human creative capacity as well as our compulsion for imitation that is the engine of human cultural adaptability. It's a "universal" of human behavior—present everywhere and insofar as one can determine, in every age. Fashion can change with breathtaking speed, reflecting neural circuits shifting to a different criterion. Consider a scene at a London ballroom at the end of the eighteenth century. Into the room filled with men in coats of blue and green satin steps a man in black, Beau Brummel. Conversation stops, everyone stares for second, but then, being gentlemen, they continue to make light conversation. But almost overnight, men's formal wear will become black.

Fashion is art for everyone. The Middle Hills and Himalayan regions of Nepal don't even have the road network of medieval Europe. Villages and isolated hamlets are linked by footpaths on which people are the pack animals. It's not unusual to see a teenaged boy or girl, or prematurely old man or woman, carrying a 100 kilogram load, supported by a tump line on the back of his or her head. Open cooking fires fill windowless huts with acrid smoke. My wife, Marcia, and I have trekked through these regions of Nepal to Inner Dolpo, an isolated pocket of Tibetan culture ringed by mountain walls. No sign of the government of Nepal was evident. There were no medical facilities, no schools, not even a trace of the Nepali police, who usually are present—their role generally being to demand and accept "inducements." Subsistence agriculture and herds of yaks and goats sustained the population. Yet even in Inner Dolpo, which can be reached only for a few summer months when the mountain passes are open, our almost obsessive human preoccupation with fashion was evident.

My wife and I walked (that's what "trekking" entails in Nepal) into Dolpo twice. The first trip in 1992 took six weeks, the second in 1997 four weeks. In both cases, we had to traverse snow- and ice-covered 18,000-foot-high passes through the mountain wall that guards Inner Dolpo. We went there to study and photograph the wall paintings of the Bon-Po religion. Bon-Po predated Buddhism, but it survives in isolated Himalayan pockets. In 1992, plastic bangles were not visible. Five years later, every woman was wearing two, three, four, or five bright plastic bangles on each arm. It's easier to understand why people would traverse dangerous mountain passes for useful things, trading salt for matches, small jugs of kerosene, and metal tools, but they, as we, are driven by fashion.

The Start of the Fashion Industry

Everyone, even the curmudgeon, is interested in fashion. The *Sunday Styles* section of the *New York Times* became popular and undoubtedly brought in a correspondingly large sum of advertising revenue, so the *Thursday Styles* section appeared. The "must-have" items featured change from week to week. That may reflect the proximate goal of the fashion industry—to convince the readers to transfer money into the pockets of the purveyors of fashion, but fashions always change. What's striking in the archaeological record is the absence of change over millions of years, until we reach the time when people who resembled us lived. And that is when the fashion industry seems to have started.

As stated earlier, works of art and fashion are one and the same and reflect the same human impulse to value essentially useless things and endeavors that do not in any manner contribute to the store of food, shelter, communication, transportation, medical wants, or comfort. Plastic bangles usually don't qualify as "mobilary art" (the jargon that you'll find in learned journals focusing on matters archaeological), but they don't differ in intent from the sea-shell necklaces fashioned by our human ancestors in Africa 124,000 years ago (McBrearty and Brooks, 2000). "Useless" works of art appear in the form of other ornaments such as pierced ostrich eggshell, bone necklaces, and pendants. Traces of red ochre used today for body painting (think of blusher), also appear, as well as complex, notched stone tools, awls, and blades. A sophisticated paint workshop existed in the Blombos Cave, about 186 miles east of Cape Town in South

Africa. The cave had been inhabited by humans 140,000 years ago. The paint workshop is dated at 100,000 years ago. Abalone seashells were used as containers to store a concoction of ochre, bone, and charcoal. The shells were reused repeatedly; the paint factory was an ongoing enterprise. The production process was complex. The paint mixtures were heated; liquefied bone marrow probably was used as a thickener. Bones were used to both stir and apply the paints (Henshilwood and d'Errico, 2011). Stone tools with inscribed decorative lines dated to 75,000 years ago also have been found in the same cave. McBrearty and Brooks (2000) in their review article trace the gradual development in Africa of the "useless" artifacts—art that differentiates us from all other species.

The European cultural revolution, about 40,000 years ago (Klein, 1999), reflects the arrival in western Europe of human colonists whose origin was African. As noted earlier, that is the case for everyone on earth—we are all African. Humans left Africa in waves over tens of thousands of years. The last prehistoric migrations probably occurred about 80,000 years ago (Templeton, 2002). And the emigrants brought their possessions with them in much the same manner as Europeans did during the colonization of the Americas. If anyone were to compare the artifacts in Plymouth, Massachusetts, in the seventeenth century with those of the fifteenth century, without knowing anything about the *Mayflower*, a "cultural revolution" would be dated to 1620 when the *Mayflower* dropped her anchor.

Someone, sometime after 260,000 years ago, the period when human beings who essentially resembled us lived, thought of grinding up soft red rocks. The selective sweep that resulted in the full form of $FOXP2_{human}$ gene had occurred (Ptak et al., 2009), and human brains probably were working in much the same manner as ours. His or her intent probably was to decorate body and/or face. Perhaps someone was smudged by a powder from a rock that had been crushed. Then the innovator was imitated, and the pigments were refined step-by-step. Most everyone coveted necklaces and painted their faces and bodies. And these "customs" and practices then were culturally transmitted, modified, and elaborated from generation to generation. Witness the painted faces of soccer fans, or the red, yellow, or saffron tikas that mark Hindu castes. Think of blushers. We all imitate; some few innovate.

At the start of each semester at Brown University, I'm tempted to ask the assembled 18- to 20-something-year-old students why they have obeyed the clothing czar and have scrapped last year's jeans, shorts, tights, whatever to conform with his latest Ukase, but I don't dare.

Neanderthals Who Did Not Imitate

Neanderthals have been an enigma since the first skull was discovered in the Neander valley. One theory was that it was the skull of a Cossack who died during the Napoleonic wars. Their brains were somewhat larger than ours. Neanderthals possessed a form of FOXP2 that differs from chimpanzees, though it's not clear whether they possessed the form that swept through humans 260,000 years ago. When humans arrived in Europe and Asia, there must have been close contact with Neanderthals as well as other, closely related archaic humans, because they mated. Advantageous genes from Neanderthals and Denisovans (a sister group to the Neanderthals; see Reich et. al., 2010) were retained in humans perhaps because they enhanced our human immune system's resistance to disease (Abi-Rached et al., 2011). Neanderthals successfully lived in cold, harsh environments, which involved using fire and clothing, and they undoubtedly possessed some form of language. They clearly were cognitively superior to present-day nonhuman primates and most likely to earlier hominims. However, the archaeological record reveals they acted in a manner unlike humans—they didn't imitate or adapt the superior technology of the humans they encountered. Imitation complements innovation, pushing forward the pace of human culture and technology. It is always the case that some few are creative innovators—the vast majority imitate, whether it is a matter of jeans or a different view of the nature of the universe, or a better way of gathering food. Neanderthals seemed to be nonimitators.

The absence of an impulse to imitate in Neanderthals shows up in other ways. The human colonists who arrived in Europe and Asia from Africa 40,000 years ago appear to have had a nomadic social organization that was better suited for hunting and food-gathering. Analyses of the teeth of the gazelles that sustained both humans and Neanderthals for a period in western Asia (the Mideast) shows, for example, that nomadic humans followed the gazelle herds. In contrast, Neanderthals inefficiently foraged from home bases and never caught on to the fact that it was far easier to follow the herds instead of dragging back carcasses to their home base (D. Lieberman, 1993). Humans may have outmaneuvered Neanderthals in other aspects of competition in the "struggle-for-existence" because Neanderthals lacked our impulse to imitate — Neanderthals were not "copy-humans."

However, some archaeologists thought that Neanderthals did, in fact, copy or perhaps even had independently invented the sophisticated stone tools associated with the modern humans who arrived in Europe 40,000 years ago. The Chatelperronian culture, based on archaeological evidence from a single cave in France, the Grotte du Renne, provided the evidence for Neanderthals copying human technology. The Chatelperronian culture ostensibly marked a transition from the cruder Middle Paleolithic tools generally associated with Neanderthals to the toolkit of anatomically modern *Homo sapiens*. (The term "Cro-Magnon," the place name where an anatomically modern human was unearthed, often is used in older studies to refer to people like us who lived in the Upper Paleolithic.) However, the Neanderthal Chatelperronian culture never existed. The evidence for the Chatelperronian culture rests on the material recovered from 15 levels in the Grotte du Renne cave. Sediment gradually piles up on buried objects over the course of time. If you find objects buried at a level lower than other objects, they usually are older. The upper level in the cave, which was dated using radiocarbon techniques in the 1950s, was attributed to humans who lived about 20,000 years ago. Radiocarbon dating is based on the percent of a radioactive isotope of carbon decays. It is an atomic clock. The lowest level in which the Chatelperronian artifacts were recovered was attributed to Neanderthals. However, newer, more precise, radiocarbon dating of the supposedly lowest Neanderthal level shows that it is actually about 20,000 years old, a date long after Neanderthals became extinct (Higham et al., 2010). The cave's levels were churned up by earthquakes and tremors. The 20,000-year-old human-made stone tools and jewelry fell down into the lowest level. Neanderthals did not copy or independently make advanced tools. Nor can any trace of art can be attributed to Neanderthals. The "Venus" statues made by humans in the Upper Paleolithic can take their place in any museum with the most highly valued contemporary art. The cave paintings of that period likewise have been equaled, but not surpassed. No trace of art marks Neanderthal culture.

Riel-Salvatore's (2010) study of artifacts from Neanderthal and human communities in southern and central Italy reinforces the view that Neanderthals did not imitate human culture or even borrow useful tools. During the transition from the Middle to Upper Paleolithic, Neanderthals lived in southern and central Italy, not far from the modern humans to their north. However, Neanderthals continued to use the same large primitive Middle Paleolithic stone tools that they had used for a hundred thousand years, taking no inspiration from the markedly better human toolkit. Virtually no change occurred in the Neanderthal toolkit. Neither creativity nor even copying occurred. That's the case even though independent studies clearly show that Neanderthals and modern humans must have been in close contact because they mated (Green et al., 2010; Abi-Rached et al., 2011).

Neanderthal conduct, in this regard, is wildly different from humans. In our travels through remote isolate parts of Asia, unreachable by road, we would always come across people walking listening to a Walkman—the miniature tape player that was then the state-of-the-art for portable music. Wristwatches were a status symbol. Steel-bladed knives, mattocks, and shovels had replaced crude home-forged iron tools. Matches and cigarette-lighters lit fires, replacing fire-bows. Plastic jugs filled with kerosene were carried over treacherous high mountain passes and snow and ice fields. For that matter, containers, especially water containers, were a valuable commodity brought in from the "outside." Virtually everyone was wearing sneakers, and the world's most ubiquitous garment, jeans, were common. Flashlights were valued items, and small gifts of batteries were warmly received. Aspirin was a miracle drug.

Back home at dinner parties, my wife and I often heard well-meaning friends lamenting the fact that "traditional" culture was disappearing as they viewed my photographs, forgetting that they, themselves were keeping up with the latest technology. Our experiences are commonplace. Traders traversed the globe over the course of recorded history, bringing goods from India to Imperial Rome. In prehistoric times, flint and seashells were transported and exchanged over thousands of miles. Why didn't Neanderthals imitate the humans with whom they came into contact? Imitation is the second engine after innovation and creativity in shaping our ever-changing culture. Why are we *homo imitatus*? Perhaps there are species-specific human transcriptional factors that we should be looking for that created *homo imitatus*?

Chapter Six

The Gene Game

Did you know that you have an insatiable desire to hang a paint-ing on your wall of a grassy field with a few trees, maybe a pool of water? It is supposed to depict the African savannah where humans might have first evolved. On a summer evening, after the NPR news, I first heard Steven Pinker claim that everyone preferred such pictures. Pinker said that surveys of people's preferences for art throughout the world showed that this was the case, though he didn't identify any of these surveys. However, the NPR interviewer was entranced by Pinker's theory.

Art

Dennis Dutton makes the same claim in his 2009 book *The Art In-stinct*, modeled on Pinker's book *The Language Instinct*. Pinker and Dutton are not talking about high art—it's instead soothing images that bring us back to the place that we all were supposed to come from—the African savannah. Pinker and Dutton both think that's where our very, very distant ancestors may have lived. Their premise is that an innate memory of the old home place is coded in the brain of every human being. Its depiction is so soothing that you want to have a painting of an African savannah, or at least a reproduction if you can't afford a painting. Unfortunately, the savannah wasn't necessarily home sweet home. Pinker and Dutton haven't kept up with paleoanthropology. Lucy and her kin, the very archaic hominims described by Don Johanson, lived in a heavily forested terrain (Johanson and Maitland, 1981).

Dutton in his book cites surveys of the art preferences of young children. The tastes of five-year-old children presumably reveal our innate art instinct before it is contaminated by memories of vacations, pet dogs, whatever. One study indeed reports that all children like to look at scenes depicting open fields with a few trees (Falk and Balling, 2010). But if you search the literature, you will find conflicting studies that state that young boys prefer seascapes, while young girls prefer pictures of people to landscapes (Dietrich and Hunnicutt, 1948; Taunton, 1982). Other studies show that all Norwegians, male and female, prefer to have seascapes on their walls, so you will be reassured that your species memory is not anomalous. These surveys should also rest your mind concerning the images on the calendars and catalogs that you surely get through the mail—they are not the products of aberrant, diseased brains directed toward persons possessing aberrant, diseased minds. One of the many catalogs that fills up our paper recycling box comes from Cabela's, the "World's Foremost Outfitter." Cabela's sells a range of useful outdoor gear and clothing as well as an array of camouflage suits that you can wear if you wish to impersonate a tree to get close to a deer. If the images of the items at Cabela's were evidence of our presumed visual species-memory, we might conclude that our distant ancestors lived in a verdant Garden of Eden full of herds of deer and moose, flocks of pheasants, streams full of fish. Cabela's catalogs have pages and pages showing framed pictures, wall-hangings, bedspreads, towels, coffee-mugs, hat racks, and toilet paper holders (virtually anything you can think of) depicting moose, deer, pheasants, grouse, trout, bass, and sundry creatures who can be eaten, as well as bears.

Cheater-Detectors

The Pinker-Dutton claim is a typical exercise in evolutionary psychology. A story is crafted about events in the very distant past, and hypothetical genes are postulated that pass on some relevant aspect of behavior to every human throughout time. Everyone now acts or thinks in some way because they have these putative genes. The masters of the genre are Leda Cosmides and John Toobey. Their 1992 book *The Adapted Mind: Evolutionary Psychology and the Generation of Culture* is a pastiche of just-so stories about how prehistoric people acted and studies of human behavior that probably have a genetic basis, such as infant-mother communication.

For example, we all supposedly have a "cheater-detector" gene, and a scenario could easily be crafted to account for natural selection favoring a selective sweep for this putative gene in which a hunter-gatherer seated before the collective fire detects someone eating more than his/her fair share of the hunt and dispatches the villain, thereby increasing the frequency of the cheater-detector gene that enabled him to detect the cheat. Scenarios that actually have been published in scholarly journals, place the evolution of the cheater-detector gene further back in time to our primate ancestors. It really is puzzling in the light of everyone now having a cheater-detector gene why thousands of investors were fleeced in Bernie Madoff's Ponzi scheme.

The evolution of language lends itself to these kinds of just so stories. Some of the stories provide relief from the tedium of academic conferences! At a UNESCO-sponsored conference in Paris in the 1980s, the room was warm and darkened and I was about to doze off, but I must have still been awake when I heard that the evolution of human speech could be traced back to an astronomical event. The speaker had said that the conjunction of Mars and Venus 4,500 years ago caused consonants to burst forth from the mouths of the startled Hebrews. The speaker was speaking English, but everyone at the conference had a set of headphones and a switch that allowed hearing a simultaneous translation into French, German, or English. I put my headphones on and switched to French to be certain that I wasn't dreaming. There is always a slight delay in "simultaneous transla-tion," but the French translation had come to a halt. After about a minute, the translator, suppressing a giggle, repeated the message.

Sex

A 2011 paper in *Nature*, which accepts less than five percent of stud-ies submitted to it, suggests to some people that human social struc-tures, particularly those governing sex, are fixed by genes that de-rive from our primate ancestors. The paper's message (Shultz et al., 2011), as conveyed to the readers of the *New York Times* by Nicholas Wade in his December 20, 2011, article, is genetic determinism. Wade reported that all nonhuman primates have the same social structures wherever they live, whether it's in a savannah, mountain-ous region, or rain forest, reflecting genetic determinism. The paper actually doesn't seem to make that claim, but Wade holds to the view that genes govern us. His 2009 book *The Faith Instinct* is based on that premise. Wade linked the *Nature* monkey and ape paper with Bernard Chapais's views on human sexual mores.

Chapais, who teaches at L'Universite de Montreal, also seems to hold to the belief that genes rule us. He states that human social structure is similar in all cultures owing to our genes, but he qual-ifies that claim by stating that culture "hides" this. The message in the *Times* article is this:

> Evolutionary change in any particular lineage is highly con-strained by the lineage's phylogenetic history. . . . This rea-soning applies to all species, including ours. But in humans, cultural variation hides both the social unity of mankind and its biological foundation.

Presumably, Chapais somehow can remove the veil of culture. Chapais then accounts for both family groups and sexual promiscuity:

> Human multifamily groups may have arisen from the gorilla-type harem structure, with many harems merging together,

or from stable breeding bonds replacing sexual promiscuity in a chimpanzee-type society.

Are we ruled by either or both chimpanzee and gorilla genes? If it's a harem, it must be a gorilla gene that rules you. If it's simply lots of sex, it must be a chimpanzee gene. Chapais's 2008 book *Primeval Kinship* apparently claims that a further stage in human social organization occurred that suppressed sexual promiscuity. Individual bands allied with those with whom they exchanged daughters. The scenario bears a striking resemblance to the marriages that bound the royal houses of eighteenth- and nineteenth-century Europe, which unfortunately did not tie the "bands" (read Germany, England, France, Austria, Russia) together and ended in the destruction of the prevailing social order in World War I. It moreover is hard to believe that anyone thinks that "sexual promiscuity" no longer characterizes human conduct. Chapais's notions of "normal" human sexual mores may reflect what he thinks *should* be going on in Montreal. But even if everyone in Montreal behaved as Chapais would have us believe, it's irrelevant. The behavior of everyone on earth does not conform to the practices of WEIRDO culture—Western, Educated, Industrialized, Rich, Democracies. As Joseph Henrich and his colleagues in their 2010 review article on WEIRDO behavior in the *Behavioral and Brain Sciences* pointed out:

> In the tropical forests of New Guinea, the Etoro believe that for a boy to achieve manhood he must ingest the semen of his elders. This is accomplished through ritualized rites of passage that require young male initiates to fellate a senior member (Herdt 1984/1993; Kelley 1980). In contrast, the nearby Kaluli maintain that male initiation is only properly done by ritually delivering the semen through the initiate's anus, not his mouth. The Etoro revile these Kaluli practices, finding them disgusting. To become a man in these societies, and eventually take a wife, every boy undergoes these initiations. Such boy-inseminating practices, which are enmeshed in rich systems of meaning and imbued with local cultural values, were not uncommon among the traditional

societies of Melanesia and Aboriginal Australia (Herdt 1984/1993), as well as in Ancient Greece and Tokugawa Japan (Henrich et al., 2010).

I doubt that either Chapais or Nicholas Wade are unaware of the range of "normal" human sexual practices throughout the world and at different epochs. Harems were accepted in the Ottoman Empire until the twentieth century. Utah entered the United States only when harems were legislated out of existence in the late nineteenth century. Harems still persist in dissenting Mormon fringe groups. Harems still remain part of the "normal" social fabric in large parts of the world. Family structure varies throughout the world. And whether someone is judged to be "promiscuous" depends on who they are. Catherine the Great, the empress who ruled Russia in the eighteenth century, had a string of lovers, but no one dared call her promiscuous. The father of her son Paul, the czar who succeeded Catherine, was her first lover.

It also is doubtful that the social structures of nonhuman primates are always uniform. Stuart and Jeanne Altmann in a 1971 study showed that baboons have very different social structures in different ecosystems. The Altmanns showed a female from a baboon troop living in a forest habitat reacting to a dominant male from a baboon group living in arid conditions. The male's "baboon-culture" licensed sex at will. That apparently wasn't the case in the female's baboon-culture. She whirled around and bit him when he attempted to mount her. Jeanne Altman's 1980 book *Baboon Mothers and Infants* demonstrated the adaptability of baboon culture. The pendulum swings. The Altmann studies are a generation removed from the authors of the 2011 *Nature* paper, *Nature*'s reviewers, and *Nature*'s editors. Genetic determinism now is in fashion. Michael Tomasello and his colleagues have shown that chimpanzees and other species are capable of learning complex tasks and interacting with humans, discerning the intentions of people with whom they are in contact by "reading" subtle cues and associating these cues with actions (Tomasello, 2003, 2009), but his insights unfortunately get less attention outside his field of study.

Creationist Linguistics

I really should have started this account of genetic invention with Noam Chomsky. The theories proposed by evolutionary psychologists derive from Chomsky's nativist views on the nature of human language.

Suppose that every human brain contained an innate store of knowledge that enabled any teenager to effortlessly attain the skills and judgment necessary to safely drive, without instruction, taking advice, or really learning anything. An innate "Universal Grammar of Driving," the UGD, an organ of the human brain, would instantiate a knowledge base concerning the acquisition of all matters pertaining to driving anywhere in the world, the explicit premise being that no one learns to drive by either observing competent drivers coping with various situations, deriving general principles from these examples, or for that matter, by heeding explicit instructions. Absent the UGD, it supposedly would not be possible for anyone to master the art of driving. The knowledge stored in the UGD simply would be activated by the teenager's exposure to local conditions. The UGD could contain a set of "Principles" such as:

A. KEEPING TO ONE SIDE OF THE ROAD
B. STOPPING AT RED LIGHTS
C. STOPPING AT CERTAIN SIGNS

The UGD also could include a set of "Parameters" that acted on these Principles so as to allow the novice to drive properly in different places. The ordered parameters would activate innate alternatives that applied to each Principle. For example, the Parameters associated with Principle A might include:

1. KEEP TO THE LEFT
2. KEEP TO THE RIGHT
1.1. PASS ON THE RIGHT
2.1. PASS ON THE LEFT
3. PASS ANYWHERE

The UGD would account for a person "acquiring" the driving technique of Massachusetts—for example, on exposure to traffic in

Cambridge, Massachusetts, the novice teenager at the wheel would instantly acquire the innate knowledge coded by Parameter 2, acting on Principle A, so as to drive on the right side of the road. Principle 2.1 then would be activated so as to convey the knowledge that s/he should pass on the left of a moving vehicle. The driver's knowledge base perhaps could then be "mathematically" described by the string (A, 2, 2.1). However, since the teenager was acquiring driving in Cambridge, Massachusetts, the epicenter for study of innate human knowledge of language, arithmetic, art, and so on, which also has high rates set for collision insurance, Parameter 3 might instead be activated. Similar Parameters and sectional constraints would mark the teenage driver's acquisition of knowledge of STOPPING—for example:

1. RIGHT TURN AFTER STOPPING
2. NO TURNS AFTER STOPPING
3. IGNORE ALL SIGNS

If a UGD existed, you might expect teenage drivers to rapidly became proficient. However, that claim also would be disputed by your insurance company.

But the actual driving records of teenagers would not refute the possibility of a UGD modeled on the theories proposed by Noam Chomsky and like-minded scholars. Their theories postulate innate knowledge and are not concerned with actual behavior, so it would be advisable to have a responsible adult, who can apply the emergency brake, sit next to the teenager at the wheel even if a UGD existed.

When the layers of jargon that envelop Chomskian linguistics are peeled away, they are more extreme than the deliberately silly case of the Universal Grammar of Driving. Noam Chomsky's views on human language reduce to the claim that the brain of every "normal" human being who ever was, lives today, or will ever be, contains an identical store of knowledge that enables him or her to rapidly and effortlessly acquire any language that has ever existed, exists, or will occur. This store of knowledge, the Universal Grammar, enables a child to "acquire" any language without tedious processes such as associative learning, imitation, or any form of learning. A child merely has to be exposed to a particular language and it will be activated in the child's brain. The UG is conceptually similar to the preloaded computer "wizards" that allow you to select English, French, German, Chinese, whatever, as the language that you will use to run an application. According to Chomsky and his adherents, who include most linguists, you did not learn to speak whatever language was used by your parents and the persons with whom you were in contact as a child. All you had to do was to hear English, or German, or Tibetan, and that triggered the correct download. Noam Chomsky's model is an improvement on any current computer software wizard that entails a mouse-click or touch-screen fingerprint to select the correct download—it is voice activated.

The hypothetical Universal Grammar is one of the key elements of Chomsky's "Faculty of Language." The concepts derive from Rene Descartes's seventeenth-century views on language, and the terms reflect those used by nineteenth-century phrenologists. As was the case for phrenology, language is instantiated in the brain by a domain-specific organ or set of organs—modules, or in recent years, domain-specific language circuits. Every "normal" human being had, has, or will have in the case of future generations, the same, identical UG. The argument that Chomsky and like-minded linguists make for every normal person's UG being identical typically starts with their asserting that every normal person has a heart, hands, and a nose. The incorrect jump in their argument is that all hearts, hands, and noses are functionally identical, which isn't the case. The false linguistic conclusion they reach is that the hypothetical Universal Grammar is functionally identical in every normal human being—this is impossible.

Darwin's Demise in Chomskian Linguistics

Chomsky must postulate an identical Universal Grammar in every dead, living, or yet-to-be-born human because the Universal Grammar is the causal agent for any human acquiring any language.

This entails every person who was ever born or will be born having the same identical UG, because any normal child can acquire any human language. As noted earlier, if variation existed on the UG, as is the case for hearts, hands, noses, or any organ of the human body, then some children would be unable to acquire a particular language because some critical feature would be missing from their UG. Some child raised in a German-speaking setting might not have the features necessary to activate German, though he or she might be able to activate Chinese. We would find thousands of children who could not acquire their native language.

The logical consequence in Chomskian linguistics is that Darwinian natural selection cannot exist because natural selection critically depends on the presence of variation. Charles Darwin in the first two chapters of *On the Origin of Species* stressed the presence of biological variation. Darwin didn't understand the genetic bases of variation, but he realized that variation is the feedstock of natural selection, as well as for artificial selection guided by agriculturalists and animal-breeders. Biological variation is an undisputed fact. Everyone does not have the same heart; everyone does not have the same hands. And human brains vary to the degree that it is often difficult to interpret the findings of fMRI brain-imaging studies.

The traditional solution taken by linguists to the problem of variation is to instead ignore it and discuss the behavior of an "ideal speaker-hearer." This approach has yielded insights on the nature of language as well as missed opportunities (some of the problems are discussed in the following and in Lieberman, 2006). Chomsky's solution is bolder; he simply sets Charles Darwin aside. As Chomsky puts it,

> It is perfectly safe to attribute this development [of innate language structures] to "natural selection," so long as we realize that there is no substance to this assertion, that it amounts to no more than a belief that there is some naturalistic explanation for these phenomena (Chomsky, 1972, p. 97).

At the 1975 Conference on the Origins and Evolution of Language and Speech, organized by the New York Academy of Sciences,

165

Chomsky repeated the preceding lines almost verbatim. He makes the same claim in his 2012 book *The Science of Language*.

Why Most Americans Cannot Speak Correct English

If Noam Chomsky were to accept the view that natural selection is the driving force of evolution, he also would find it impossible to account for why he and most Americans can speak English. As chapter 3 pointed out, natural selection on humans never ended. Natural selection will shape any aspect of human biology that contributes to survival of an individual and his or her children. The ability to digest cows' milk as an adult isn't learned—it entails having an innate, genetically transmitted, biological mechanism. In human groups who possessed herds of animals that could be milked, natural selection acted to yield individuals who had the genes that enabled them to digest milk as adults, thereby enhancing their survival and their children's survival. Natural selection acted on different genes to confer the ability to use milk as an additional food source in different parts of the world at different times (Tishkoff et al., 2007). Other human groups who didn't have domesticated milk-producing animals didn't evolve adult lactose tolerance—that's why lactose-free food is stocked on the shelves of your supermarket.

While you can survive without drinking milk as an adult, without language you would be nothing. No object used in daily life, your clothing, your home, anything resulted from an innate mechanism, nor was the manner in which it is used innately transmitted to you. That is also the case for aspects of human behavior. You didn't have a gene that caused you to look both ways before you crossed the street, or a gene that enabled you to lace your shoes. Virtually all aspects of daily life are culturally transmitted, the result of cultural aggregation and transmission. And the medium of transmission is human language. Thus if your ancestors had been using any particular language for an extended period—let's say the last three thousand years (the period when adult lactose tolerance or high-altitude adaptation evolved by means of natural selection)—then if any innate Universal Grammar (UG), or Optimality Theory (OT), or other variant of Chomsky's Faculty of Language were *necessary* to "acquire" that language, natural selection would have acted to optimize it.

The United States is an optimal experiment-in-nature, since the ancestors of most Americans did not speak English before they arrived here. Because languages differ dramatically, you might have never been able to acquire English if your ancestors were speaking Chinese, Hungarian, Italian, and so on. Your UG would have been optimized for your ancestral language, rendering it useless or deficient for acquiring English. Only Americans whose ancestral lineage was English—which includes neither Chomsky nor me— would be able to speak "correct" English. As chapter 3 pointed out, Americans having Chinese ancestors surely would not be able to even master the syllable structure of English because the time depth of the Chinese languages exceeds the period in which natural selection yielded adult lactose tolerance, or the respiratory capacity to adapt to extreme altitude.

Back to the 1960s

Chomsky's views on innateness and language seem to be an amalgam of Rene Descartes's seventeenth-century notions of Universal Grammar (which was supposed to prove the existence of the soul) and the findings of twentieth-century studies on how short-lived animals' genetically transmitted behavior is triggered by the environment. He still tenaciously holds to the view that similar mechanisms account for how children acquire language. Geoffrey Pullum, who heads the Language and Linguistics program at Edinburgh University, was surprised to hear Chomsky once again repeat his tale of the rock, rabbit, and granddaughter at a conference that took place in October 2011 in London to an audience of invited academics. The tale, which appears in several of Chomsky's publications, is

> To say that "language is not innate" is to say that there is no difference between my granddaughter, a rock and a rabbit.

> In other words, if you take a rock, a rabbit and my grand-
> daughter and put them in a community where people are
> talking English, they'll all learn English. If people believe
> that, then they believe that language is not innate. If they
> believe that there is a difference between my granddaugh-
> ter, a rabbit and a rock, then they believe that language is
> innate (Chomsky, 2000, p. 50).

The tale of the rock, rabbit, and granddaughter would make sense
if "duckling" were substituted for "granddaughter" and "quacks" for
language.

The experimental data underlying Chomsky's views on how children acquire language appears to devolve from a 1960s debate on the
nature of bird brains. Psychologists then were debating the degree to
which animal behavior was malleable. One point of view was that
brains are very malleable, and "imprinting" experiments supported
that view. Duck or chicken eggs would be incubated in a laboratory
isolated from adult ducks or chickens. The newly hatched chick
or duckling, if exposed to a person, would follow that person about
as though he or she was a mother duck or hen. The premise was
that during a "sensitive" period after birth, the bird brain was completely malleable and would take the first object or person it saw as
its mother. Gilbert Gottlieb was a young psychologist who wondered
whether this malleability extended to everything birdlike. It didn't.

Gottlieb and his friends found that mallard ducklings did not respond to wood duck maternal calls, nor did wood duck ducklings
respond to the maternal calls made by mallard ducks. In a brilliantly
executed project reported in a 1975 paper, Gottlieb showed that
ducks didn't have to learn everything needed to be a duck. Duck
information was preloaded in their brains that merely had to be triggered. When a duck egg that was about to hatch was briefly exposed
to recordings of adult ducks quacking and was then kept isolated
from other ducks, months later the duck would quack appropriately.
When duckling eggs were not exposed to quacks, the grown ducks
never quacked like a duck. There is a "sensitive" period that starts
shortly before a duck is hatched in which a duckling in its egg, or

shortly thereafter, merely has to hear duck calls for it to acquire the calls. This makes sense for animals that have a short life span. They don't have to learn to quack, nor do they probably have to learn other things that are necessary to be successful in duck-life. There is an innate, domain-specific "Faculty of Quacking" in the brain of every mallard duck, wood duck, or other type of duck. But that doesn't work for children — try to get your child to speak Tibetan by playing a CD of someone speaking Tibetan shortly after she or he is born.

Chomskian linguistics purports that if baby Susan or Sam hears English, French, or whatever language is being spoken, she or he will master English, French, and so on by age four or five years. That doesn't happen; children don't rapidly master either speech or syntax. Carol Chomsky in her 1969 book found that English-speaking children were not up to speed at age eight years. Robin Burling's 2002 review article confirms that French- and English-speaking children haven't mastered the syntax of either language until they reach their teens. Children don't even talk as fast as adults or achieve adult articulatory proficiency until their teens (Smith, 1978; Smith and Goffman, 1988). Melissa Bowerman showed that Dutch children who heard hours of German every night on the TV programs that their parents watched neither spoke or understood German until they later studied German in school. Children learn to talk and master a language by using that language over a period of at least a decade.

Another questionable thesis of Chomskian linguistics is the "poverty of the stimulus." Accordingly, absent UG, children would never master the syntax of their native language, unless they were overtly corrected whenever they uttered a grammatically incorrect sentence or phrase. Mommy, Daddy, or someone else, would have to continually correct them. That usually doesn't happen; hence, Chomsky concluded that there *must* be a UG in every child's brain. A mathematical exercise, Gold's theorem (1967), is the theoretical basis for the poverty of the stimulus argument. Gold's exercise showed that "negative" information, explicit correction of a child's errors, is necessary for *perfect* transmission of the syntax of a language. Apart from the fact that children are indirectly corrected when they make mistakes—adults or older children will often immediately echo a young child's garbled utterance, correcting the error—Gold's theorem was shown to be irrelevant two years later when Horning, using similar mathematical methods, showed that negative information isn't necessary if less than perfect correctness is attained. One of the mechanisms for language change is the fact that children do not exactly copy their caretaker's speech. Horning's Stanford University dissertation is accepted by specialists in computer modeling and mathematical logic. It generally is ignored in arguments for an innate Universal Grammar.

The 1957 Starting Point

None of this nonsense was proposed in Noam Chomsky's first class at MIT. I was one of the four students. It probably was the fall of 1954, perhaps 1955—I don't have a transcript of my MIT grades— and the young Noam Chomsky said nothing about evolution or innate faculties of the mind—*Cartesian Linguistics* was years ahead. Chomsky's 1957 book *Syntactic Structures* had not yet been printed. Instead, we used a blue Ditto-Machine version! If I had preserved it, it would now be a collector's item.

Chomsky claimed that in your mind/brain, the meaning of a sentence such as *I saw the boy who ate ice cream fall down* was represented by an "underlying" structure in which the simple "canonical" sentence *The boy ate ice cream* had been inserted into the framework of the canonical sentence *I saw the boy fall down*. The underlying representation was supposed to reflect how your mind discerned the meaning of the sentence. "Transformational" rules then acted on the underlying representation to produce the complex "surface" sentence, *I saw the boy who ate ice cream fall down*, that you spoke or wrote. Innate organs of the brain and evolution were never discussed. The theory presented in *Syntactic Structures* was elegant and Noam was a modest, charming teacher. I didn't begin to see the problems until years later, when in MIT's graduate program in linguistics it became evident that the underlying representation of a sentence of any language whatsoever, always was a simple English sentence. English apparently was the "universal" language of thought.

I also came to wonder about the ultimate objective of linguistic research. Linguists almost always attempt to describe the behavior of an "ideal speaker-hearer." Variation is ignored or treated as "noise" that detracts attention from the true knowledge of language—the competence—that everyone surely possesses. The objective then becomes one of discovering "laws" of language, similar to the laws of Newtonian physics that govern the formation of sounds, words, and sentences—in short, a model that stands in stark contrast to what Ernst Mayr (1982) termed the "population thinking" of biology, where variation must be taken into account. The "laws" of Newtonian physics work because all rocks will fall at the same rate. In contrast, Chomsky's grandchildren will reach different adult heights, even if they were fed the same food and held to a similar regimen in all aspects of childhood life. If you have ever read the information packed with a prescribed medicine, the range of "adverse" effects that can occur should convince you that people vary with respect to their reactions. Some few linguists have taken account of variation, such as William Labov (1972), who documented the manner in which dialects change over time. However, in most linguistic exercises, any unwanted deviation from the "competence" predicted by the theory in question is dismissed as a "performance" effect. Just how anyone decides what competence is remains a dark art, and it usually turns out to be behavior that supports someone's pet theory.

Chomsky and linguists sharing his views have over the years devised ever more complex and opaque jargon, but the basic principle remains unchanged. They're the same in his most recent publication (Chomsky, 2012b). The complex, grammatical, well-formed, sentences that you might hear or read are generated from a simple sentence into which other sentences have been inserted. The simple sentence is the object for semantic interpretation—what the sentence means. The sentence that you hear or read has been modified to match "sensory motor" constraints (Chomsky, 2012a,

p. 13). I was studying electrical engineering at MIT when I first encountered Chomsky's "formal linguistics," which involved writing "rules of grammar" that are similar to the algorithms then used by software engineers to write computer programs. Whereas the art of software engineering has made remarkable progress since 1957, virtually no progress has been made even in describing English, the most intensively studied language on earth. Linguists continue to use algorithms that haven't changed from the ones we used in Chomsky's first class at MIT. Even linguists holding to Chomsky's general conception of the nature of human language, such as Ray Jackendoff, admit that

> Thousands of linguists throughout the world have been trying for decades to figure out the principles behind the grammatical patterns of various languages. . . . But any linguist will tell you that we are nowhere near a complete account of the mental grammar of any language (Jackendoff, 1994, p. 26).

The Key to Human Language, According to Chomsky

Chomsky has never abandoned the early version of Universal Grammar (UG), which had hundreds of innate "principles and parameters" that supposedly were preloaded into every human brain. The 1980 version of UG, with its large inventory of principles and parameters, seemingly disappeared in 2002, replaced by two Faculties of Language. However, UG has reappeared in his subsequent publications, such as his 2012 paper "Minimal Recursion." Optimality Theory, which attempts to account for the manner in which children "acquire" the speech sounds and syllabic structure, also involves a battery of principles and parameters.

The "Faculty of Language, Narrow" (FLn) provides "recursion" — the hypothetical, defining core feature of human language (Hauser et al., 2002). The "Faculty of Language, Broad" (FLb) transmits those elements of language that are shared with other aspects of human behavior. Recursion, the FLn, boils down to being able to construct sentences that include clauses, like this sentence — or sentences such as *I saw the boy who ate an ice cream cone fall down.* Or even sentences that young children often form such as, *I saw the elephant and the hippopotamus at the zoo.* Chomsky's core constituency now takes recursion to be the defining feature of human language. Hauser et al. (2002), the initial paper on recursion, stated that recursion was domain-specific, occurring in language and language alone. No other living species supposedly possessed recursion, which in their view distinguished language from all other aspects of human behavior. In subsequent publications and talks, Chomsky with Marc Hauser and W. Tecumseh Fitch have defended this view of language with slight modifications (for example, Fitch et al., 2005; Fitch, 2010). Unfortunately, Daniel Everett came along and showed that a language spoken in the depths of the Amazonian rain forest, Piraha, does not have sentences that involve recursion.

The argument for recursion being the key to language is odd because virtually every aspect of human and animal behavior involves recursion. Think of a square dance or the more elaborate dances of the nineteenth century — a video of a dramatization of any of Jane Austen's novels will demonstrate a couple holding each other as they swing around, moving in the path formed by a line of men and women to its head. Their dance steps are then repeated by the next couple. Recursive acts are inserted into other elements of the dance. Virtually all musical forms entail recursion. Walking involves recursively inserting movements such as heel strike into a larger pattern. If the syntactic "rules" used to describe sentence structure were used to record the actions taken by squirrels stashing nuts away for the winter, the squirrel syntax would involve recursion. As a squirrel dashed about looking for places to bury a nut, s/he would recursively perform the same sub-sequence — digging a hole, placing the nut within, and then covering it. Each nut burial constitutes a recursive act, placed within the framework of the squirrel's dashing about, searching for suitable locations. The stereotyped innate grooming patterns of rats and squirrels are regulated by recursive neural firing patterns in their basal ganglia (Aldridge and Berridge, 1998). The grooming patterns involve rats performing a sequence of paw and body movements in complex sequences that include nested sub-sequences. Wayne Aldridge and Kent Berridge kindly sent me a video of a California ground squirrel performing his charming (to Ms. Squirrel) courtship dance. Mr. Squirrel whirls around, while his tail recursively does all sorts of things that will entice Ms. Squirrel to be his partner.

Birds can learn to recognize birdsongs that have different recursive patterns. In a study published in *Nature* in 2006, Timothy Gentner and his colleagues at the University of Chicago showed that starlings could recognize acoustic patterns that were differentiated solely by their recursive patterns. Though most people don't think of starlings as "songbirds," they sing long songs that are composed of smaller iterated "motifs"—acoustic segments that together constitute songs that the birds recognize. Eleven European starlings (*Stuns vulgaris*) were trained to classify sub-sequences of songs formed from starling motifs by grammars that involved recursion, as well as grammars that didn't. The grammatical algorithms were similar in nature to those that linguists use to describe sentences. Nine of the starlings were successful in identifying birdsongs generated by recursion. Curiously, Chomsky's published response to the starlings being able to learn these recursive syntactic rules was that it was the result of natural selection, which according to him, doesn't account for the evolution of the neural bases or anatomy involved in human language and speech.

Amen Brothers, Amen Sisters!

I missed a golden opportunity at the plenary session of the 1995 Boston University Conference on Language Development when an apostate recanted. I failed to rise up, throw my arms upward and declare, "Hallelujah! Amen Sisters, Amen Brothers."

Jill de Villiers, who for decades had studied how children learn to talk and master English syntax, declared that she now "believed in Universal Grammar." Almost everyone applauded. That wasn't the first or last time that I heard a linguist declare, "I believe in Universal Grammar."

A theory, in the domain of science, cannot be "proved." You don't have a theory unless it can be tested—subjected to falsification. A belief is inherently different. How can anyone show that heaven does not exist? You can believe in the Holy Trinity, the Immaculate Conception, Moses receiving the Ten Commandments, Tibetan deities, or other metaphysical entities. Belief rests in the domain of religion, not science. When Evangelical fundamentalists argue against Darwin, they often assert that the theory of evolution is "just" a theory, because it can't be proved. That's what it is, because a scientific theory cannot be proved—it, however, *must* be formulated so that it can be falsified.

Chomsky doesn't really have a theory, because it cannot be falsified. He has repeatedly stated that the Faculty of Language specifies a speaker-hearer's *knowledge* of grammar; it does not predict what s/he actually understands, says, writes, signs, or texts. When Nicholas Evans and Stephen Levinson in their paper, "The Myth of Language Universals," pointed out that they could not find any evidence for the presence of the syntactic "language universals" that Universal Grammar ostensibly transmits, it had little effect. Since the Universal Grammar, by definition, transmits the features of all language, the features of any language or the absence of any feature fit its "predictions." In short, the predictions of Universal Grammar cannot be falsified because it predicts nothing. It's a matter of faith. That also turns out to be the case for recursion. In Chomsky's 2012 essay "Minimal Recursion: Exploring the Prospects," he again takes the same path, making it impossible to refute the claim that recursion is the defining feature of human language. On the first page of his essay, Chomsky states that the "range" of the "recursive procedure" could be "null." Thus, though a language might lack recursion, seemingly refuting recursion being the defining property of human language, it wouldn't matter. This apparently is Chomsky's way of dealing with Daniel Everett's finding that the language of the Piraha people lacks recursion. Everett's groundbreaking study, whose message is far broader than recursion or its absence, will be discussed in the next, concluding chapter.

A Minor Miracle

Every faith needs at least one miracle. Nicaraguan Sign Language (NSL) fulfilled that end. Judy Kegl, an MIT-trained linguist, visited Nicaragua shortly after the Sandinistas had conquered the country. The Sandinistas had overthrown the right-wing Somoza dictatorship. Kegl's Spanish was elementary, and so she relied on her Sandinista guides. The story that Kegl brought back was that deaf children in Nicaragua had been totally neglected by the former Somoza government. The new Sandinista government gathered the deaf children. A trained sign-language instructor was brought in from the Soviet Union, but the instructor became irrelevant. When the children were gathered together, they spontaneously developed a complex sign language without benefit of the Soviet teacher. The published paper (Kegl et al., 1999) claimed that the spontaneously created sign language, NSL, conformed to the principles of Chomsky's Universal Grammar. Magazine articles, documentary films, talks at major conferences, and so on publicized Kegl's NSL story. It lives on in Chomsky's circle of like-minded scholars.

The miracle story lasted seven years. When Laura Polich, who specializes in deaf education and is a fluent speaker of Spanish and sign-language expert, visited Nicaragua, she discovered that rather different events had occurred. The former dictator's wife had sponsored a school for the deaf. An American Peace Corps volunteer had taught American Sign Language at that school, which had close contacts with schools for the deaf in Costa Rica. And the child who had played a central role in the development of NSL had attended a school for the deaf in Spain for eight years. Virtually all the information related by Kegl was wrong; the details are in Polich's 2006 book, whose focus is on the benefits of sign-language instruction to the deaf.

Nicholas Wade swallowed Kegl's story hook, line, and sinker. His September 4, 2004, *New York Times* article states that "a hearing person, miming a story about a cat . . . will make a simple gesture . . . but the deaf [Nigaraguan] children have developed two different signs to use in its place." This supposedly proved to Wade and Steven Pinker, whom Wade quotes, that NSL exhibits the properties of Chomsky's Universal Grammar in that it captures a principal quality of human language—"discrete elements usable in different combinations." By this measure, Alan and Trixie Gardner's home-raised chimpanzee Washoe who learned about 150 words of American Sign Language exhibited the principles of Universal Grammar when she described a duck using the ASL sign "water" followed by the ASL sign "bird."

Our Moral Gene

At the start of this book, I discussed Sam Harris's fMRI study, which aimed at isolating the human brain's center of religious belief. Genes that confer morality and religion are a hot topic. The title of Marc Hauser's book, *Moral Minds*, might lead you to believe that it was about moral conduct. Hauser starts off more or less on Mount Sinai with a Twitter-like version of Yahweh's commandments. Hauser doesn't mention worshipping false gods or keeping the Sabbath, but the rest is what you might have heard in Sunday School:

> All of the following actions are universally forbidden; killing, causing pain, stealing, cheating, lying, breaking promises, and committing adultery (Hauser, 2006, p. 48).

However, if you read on into *Moral Minds*, you will discover that Hauser isn't attempting to account for what people actually do or should do. "Knowledge" of morality is Hauser's concern, and that leads to Hauser providing a free pass for any kind of behavior. Mahatma Gandhi, Gengis Khan, Jesus, the Buddha, Adolf Hitler, all acted morally—take your pick.

Hauser's views on the nature of human morality are modeled on the Faculty of Language, which imparts "knowledge" of the syntax of all languages, ostensibly allowing anyone to "acquire" any language. In Hauser's case, his "Universal Moral Grammar" licenses any mode of conduct. Page 53 of *Moral Minds* sets forth Hauser's model, including the "ten principles" that constitute the "road map" for the rest of the book. I've omitted two principles in the following excerpt, but you can see where Hauser's road leads to:

1. The moral faculty consists of a set of principles that guide our judgments but do not strictly determine how we act. The principles constitute the universal moral grammar, a signature of the species.
2. Each principle generates an automatic and rapid judgment concerning whether an act or event is morally permissible, obligatory, or forbidden.
3. The principles are inaccessible to conscious awareness.
5. The principles of the universal moral grammar are innate.
6. Acquiring the native moral system is fast and effortless, requiring little or no instruction. Experience with the native moral system sets a series of parameters, giving rise to a specific moral system.
7. The moral faculty constrains the range of possible and stable ethical systems.
8. Only the principles of our universal moral grammar are uniquely human and unique to the moral faculty.
10. Because the moral faculty relies on specialized brain systems, damage to these brain systems can lead to selective deficits in moral judgments. Damage to areas involved in supporting the moral faculty (e.g, emotions, memory) can lead to deficits in moral action—of what individuals actually do, as distinct from what they think someone else should or would do. (Hauser, 2006, p. 53).

If you didn't get Hauser's message by page 53, it becomes clear when you reach page 300 that anything goes:

The universal moral grammar is a theory about the suite of principles and parameters that enable humans to build moral systems. It is a toolkit for building a variety of different moral systems as distinct from one in particular. The grammar or set of principles is fixed, but the output is limitless within range of logical possibilities. Cross-cultural variation is expected as does not count as evidence against [it] (Hauser, 2006, p. 300).

One example of conduct sanctioned by universal moral grammar (pp. 111–112) is "honor killing." Honor killing licenses brothers and fathers murdering a sister or daughter to preserve a twisted notion of family honor. The opening pages of chapter 4 of *Moral Minds* lists other examples of societies in which incest or causing pain to children to amuse yourself are "morally" acceptable. But despite Hauser's attempting to craft a theory that is irrefutable, the lessons of history demonstrate that morality doesn't hinge on genes.

Viking "Moral" Conduct and Viking Genes

Iceland is a stunningly beautiful place, and Icelanders are very friendly and helpful. But Iceland would not have been a vacation destination a thousand years ago. The Icelandic sagas record actual events. If you visit Iceland, you can stay overnight at a farmhouse close to where the events in *Njal's Saga, The Story of Burnt Njal* unfolded. The English translation by Sir George W. DaSent starts with

There was a man named Mord whose surname was Fiddle; he was the son of Sigvat the Red, and he dwelt at the "Vale" in the Rangrivervales.

When my wife, Marcia, and I went to Iceland, we were able to find the stream at which heads were split open, and where the other bloody events related in the Saga took place about 1000 AD. You can walk over the fields where spears were thrust into the backs of a father and son tilling their land and where homes were set on fire so as to roast everyone inside. There were laws and lawyers in Iceland. If you acknowledged killing anyone at the Althing, the periodic assembly at which the law was read aloud, all was right. In this setting, similar means could be used to supplement one's income during the summer months. Njal's saga relates two farmers chatting and deciding that the season is now right for raiding. Nordic farmers and fishermen at their summer jobs were the Vikings who pillaged England and Ireland, looting, killing, raping—and bringing slaves back home.

My wife is the local coordinator for Amnesty International (AI). AI published a thick book each year that noted human rights abuses throughout the world. Iceland and Norway—the Viking home bases—are the only two countries that consistently have no entries. Iceland retained the Norse custom by which Erik's sons bear the last name Eriksson, while Erik's daughters' last name is Eriksdottir. One of our Icelandic graduate students can trace all of her ancestors back to the year 1000, because genealogical records extending back to the year 1000 keep track of who is related to whom. That's true for most other Icelanders and accounts in part for Iceland now being a center for research on human genetics. The other factor is an extremely high level of education and concurrent technical capabilities.

The Viking raids ended after 1030 AD, when Christianity came to Iceland. The genealogical records show that the gene pool of Iceland hasn't changed since Viking days. Present-day Icelandic moral "performance," what Icelanders actually do, is the polar opposite of Viking times. If any change in anyone's "knowledge" of morality occurred in Iceland, it didn't involve genes, or for that matter any aspect of biology.

A Free Pass for Genocide

Apart from its problems as scientific theory, Hauser's claim for a genetic basis for morality provides a free pass for genocide. Germany during the Weimar Republic had a uniformed police force, the Ordnungspolizei. Anti-Semitism was entrenched in European culture, and Germany was no exception. However, a properly dressed family strolling in one of the Weimar Republic's many parks in 1932 would have been met with courtesy by a policeman if they had any reason to request assistance or report any untoward occurrence—even if they were Jewish. Eight years later, that same policeman might have been positioning a three-year-old Jewish girl, who had been stripped of her clothes, on a board positioned over the pit in which the bodies of her mother, father, brother, and older sisters lay in pools of blood. He then would have placed his police-issue pistol against her head and pulled the trigger.

As the German army advanced into Poland in 1939, six Einsatz-gruppen der Sicherheitspolizei followed. Their mission was arresting left-wing activists, confiscating weapons, and murdering Jews. The Einsatzgruppen used simple methods. Captive Jews were forced to dig pits, then stripped naked; they were placed one-by-one onto boards positioned above the pit, and shot in the head. The Jews instead might simply be lined up at the edge of the pit before they were murdered. But Germany was short of troops, and so as Nazi armies advanced into Russia, ten police battalions were formed to further the work of the Einsatzgruppen.

The members of the police battalions were the "ordinary" German Ordnungspolizei who had kept order and patrolled the streets of Germany (Westermann, 2005).The Ordnungspolizei diligently did their duty and murdered about one million men, women, and children. However, it became apparent that they were under stress. Their police-issue pistols had to be reloaded after eight shots. This might have caused some degree of distress when the ordinary policeman at his ordinary daily job looked into the eyes of the three-year-old girl staring at him. She might have brought up the image of another three-year-old somewhere else at some other time. But it would be necessary to put the pistol against her head and pull the trigger. Nazi records note that individual members of the police battalions often asked for other assignments, but the battalions remained at work in Russia until 1944. Mechanized death camps using gas chambers and incinerators provided a more efficient means for mass murder and genocide and also harvested silver and gold fillings from the gassed Jews' teeth and wedding rings, thereby offsetting operating expenses. Lampshades made of human skin provided arts and crafts diversion to the SS camp staff.

Selective sweeps can result in a mutation becoming established when it leads to a reproductive advantage. That was the case for lactose tolerance for people who kept herds, oxygen transfer at extreme altitudes for Tibetans, and resistance to endemic diseases in different human groups. However, no selective sweep replaced any German moral genes in the short Nazi era, nor did the moral gene return in 1945 when it became quite clear that Adolf Hitler had led Germany into the gates of Hell. The members of the police battalions were the Ordnungspolizei who had before the Nazi era issued traffic tickets.

Morality "Research," Toy Experiments, and WEIRDOs

The June 25, 2010, issue of *Science* commented on the weird view of human nature that follows from running experiments on WEIRDOs—"people from Western, educated, industrialized, rich, and democratic cultures" (Jones, 2010). A whopping 96 percent of the subjects who participated in all of the empirical studies in six top-tier psychology journals between 2003 and 2007 were WEIRDOs (Arnett, 2008). The experiments explored phenomena as diverse as visual illusions and moral judgments. A review article published in the *Behavioral and Brain Sciences* concurred (Henrich et. al., 2010). I noted some of the accepted, "normal," sexual practices of some non-WEIRDO peoples that were cited by Henrich and his colleagues earlier, when discussing theories on the genetic bases of marriage, promiscuity, and harems.

WEIRDOs also were the subjects who played the highly constrained mind-games that form the empirical data for *Moral Minds*. WEIRDO responses on the thought experiment called the "trolley" dilemma and similar exercises were supposed to reveal the psychological and biological bases of moral judgments. The trolley experiment is a "thought-experiment"—some would say a "toy" experiment—in which a subject is asked whether he or she would save five or so people who would die on a runaway trolley car that would otherwise crash, by diverting it onto another track, where it will come to a safe halt. But the cost of saving five people would be mowing down an innocent person who happens to be crossing that track. The numbers of people in the trolley and innocent people sacrificed can be varied. One of the key elements of Hauser's "moral" Universal Moral Grammar has to do with how the runaway trolley is diverted onto the track that will save its passengers. When WEIRDO subjects are asked if they will push a button that mechanically throws a switch to divert the trolley they say "yes." If the WEIRDOs are taking the Moral Sense Test, a computer game on the Harvard University website, they will mouse-click "yes." However, when WEIRDOs are asked if they would divert the trolley by pushing someone into its path, their response is "no." Physical contact thus is taken to be a universal moral no-no. Hauser correctly states that he has run thousands of subjects on this and variants of the trolley problem who respond in this manner (p. 32, n. 26). You too can join the thousands of subjects. However, to take the Moral Sense Test, you must have decidedly WEIRDO qualifications—access to a computer linked to the Internet.

Marc Hauser no longer is a member of the Harvard University faculty; he resigned from his tenured professorship in August 2010 after an inquiry concerning the experimental data in some of his published papers, including some of the monkey studies discussed in *Moral Minds*. However, the Moral Sense Test is still is on the Internet, at http://moral.wjh.harvard.edu. The test consists of a set of short paragraphs that present "moral" dilemmas involving saving some persons at the cost of killing others, or performing some act that might be acceptable, but involves harming innocent people. Other scripts attempt to assess whether failing to act is as reprehensible as an overt act. One script asks whether it is moral to turn a lifeboat away from one drowning person so that you can instead rescue five persons who otherwise would drown.

The fMRI neuroimaging study by Shenhav and Greene (2010), who worked with Marc Hauser, monitored subjects while they judged the moral acceptability of the scripts on the Moral Sense Test. Their findings show that the Moral Sense dilemmas elicit neural activity in the regions of the brain involved in executive control— the prefrontal-basal ganglia circuits that were discussed in chapters

2 and 4. Shenhav and Greene conclude that the moral judgments engage a "domain-general" cognitive network rather than anything specific to morality. Whether the Moral Sense Test actually taps the cognitive decisions that guide moral actions in anyone's life is another question. In the few real-life situations that I know of where similar decisions had to be made, there was no time to calculate the number of casualties versus saved people.

Other studies using WEIRDO subjects also conclude that "personal force" — coming into contact with someone — precludes pushing or shoving someone to save someone, and inhibits that action (for example, Greene et al., 2009). Yet the cadres of the Khmer Rouge who tortured and murdered their countrymen in almost unimaginably brutal ways would not have hesitated to push someone into the path of a truck. When I visited Cambodia, I learned that the "standard" way of killing a family was to start with the infants, taking them by their feet and swinging them so as to smash their skulls against the side of a tree in full view of their siblings, mother, and father. The mother then would be gang-raped and disemboweled, and then they would move on to kill her husband, by means that I would just as soon not describe. If we accept Hauser's position that these acts were moral, the moral code of any culture is always "moral" because it has been formed by activating the innate "principles and parameters" of Hauser's innate "Universal Moral Grammar."

The Universal Moral Grammar also didn't seem to be universal when Marc Hauser teamed up Linda Abarbanell, a Harvard University anthropologist whose field of expertise is Mayan peasant life in traditional *parajes*, small, relatively isolated hamlets in the highland region of Chiapas, Mexico. The Tseltal Mayans who were studied are the descendants of the Mayans of the classic civilization that crashed about 900 AD, leaving behind cities and temples abandoned to the jungle. The Tseltal Mayans practice slash-and-burn subsistence agriculture and most are illiterate. They almost never leave their *paraje*, where everyone knows everyone else and what they are doing. Traditional beliefs and religious practices have been preserved. In short, life is quite different than in Harvard Square. The Mayans' responses to the puzzles on the Moral Sense Test, translated into the Tseltal language, also were different from those reached by the indigenous population of Harvard Square, or the virtual population on the Harvard University website. Physical contact played no role in Tseltal Mayan moral judgments. They had no problem pushing someone into the path of a truck to save its passengers from death. They also did not distinguish between harm inflicted on a person intentionally or through omission, such as killing someone or letting someone die through inaction. Both scripts elicited approbation.

The Abarbanell and Hauser (2010) paper has a long explanation of why Mayans judge killing someone by a deliberate action or through inaction equally wrong. Abarbanell and Hauser call attention to the cultural differences between WEIRDOs (they don't use that term) and the Tseltal Mayan's close-knit *paraje* community life. However, not a word is expended to attempt to explain why direct physical contact plays no role whatsoever in Mayan moral life.

The Selfish Gene

Richard Dawkins, who works within the framework of evolutionary biology, doesn't owe anything to Noam Chomsky. Dawkins also is a forceful advocate of atheism, unlike Darwin, who kept his doubts on religion in the closet. However, Dawkins has introduced an element of quasi-religious mysticism in his gene-centered view of selection.

According to Dawkins, evolution acts to maximize the transmission of the total inventory of an individual's genes to a new generation rather than as Darwin claimed, the individual and his/her offspring. Since any individual shares genes with a group of related individuals, evolution thus would act to maximize the transfer of the genes shared by the group. That may involve an individual acting for the greater good of the group, rather than himself or herself—in short, altruism. It also follows that the more two individuals are genetically related, the more they will act to favor each other. One objective of Dawkins's 1976 book *The Selfish Gene* thus was to provide a biological basis for altruism.

185

Altruism doesn't immediately seem to fit the Darwinian paradigm. Why should an individual act to further some other individual or group in the struggle for existence? If you accept the teachings of many religions, then altruism makes sense. Altruistic conduct will add points to your Karma if you accept the teachings of the Buddha. It might get you a ticket into Heaven or at least avoid Hell if you accept the teachings of the varieties of Christianity that stress good works. However, if you are attempting to explain how we act without invoking religious belief or cultural mores, then some other explanation is necessary. The "scratch-my-back" theory is often proposed to account for humans acting in an altruistic manner. I'll do something for you, and you will probably reciprocate, but this scenario seems unlikely for many animals, insects and plants. In the 1960s, V. C. Wynne-Edwards (1962) introduced the concept of "group selection" to account for selection that seemingly was "for the good of the species" rather than any individual. Altruism in this sense applies to organisms that didn't think. Any evolved behavior that involved a cost that did not enhance the survival of the individual's offspring (seedlings, nuts, and so on, for plants) was considered to be "altruistic" or "cooperative." Dawkins's gene-centered theory can be regarded as an attempt to account for altruism without invoking group selection.

Darwin was wrestling with the problem of accounting for the "transmutation of species" when he hit upon natural selection. Natural selection is blind and acts absent any directing entity. Darwin's notebooks and letters reveal that he foresaw that he would be accused of destroying belief in God. God is not a necessary agent in Darwin's worldview. Dawkins is a "public" atheist, arguing against religious belief in any shape or form, but Dawkins has replaced God with the Gene as the directing force for the "transmutation" of species.

What Happens to Your Children — Testing the Selfish Gene Theory

If we take Dawkins's selfish-gene theory at face value, then we should always favor our children in all circumstances. Apart from your siblings, your children's genes are most likely to share some of

186

your genes. Though we often may be at odds with our siblings, children tend to be favored—that propensity may be one of the aspects of human behavior that has a genetic basis. The succession rights of monarchies are ordered on this ground, although women are usually excluded and brothers are sometimes smothered, stabbed, poisoned, or imprisoned when succession is in doubt. But setting apart from machinations of royalty or wealth, it is clear that culture overrides any selfish gene or genes.

Consider Sally Hemings and her children, who as I noted earlier also were Thomas Jefferson's children. Sally (Sarah) Hemings was in Jefferson's lifetime said to be his mistress, but the term "mistress" doesn't quite seem to fit a woman who has been your companion for 38 years and has borne six of your children. In contemporary accounts, Sarah Hemings was said to be "mighty near white . . . very handsome, long straight hair down her back." That wasn't surprising since she was the daughter of John Wayles, who was the father of Jefferson's "white" wife. Wayles also had five other children with his "black" slave Betty Hemings. The quotation marks on "black" are appropriate because Betty Hemings also had a white father, John Hemings. Her mother was John Hemings's slave Susannah. Most African-Americans have "white" ancestors because slave-owning fathers commonly sold their daughters and sons into slavery. The laws of Virginia treated the children of slaves as slaves.

If not for the corrupting slave-owning culture and laws of Virginia, Jefferson could have married Sally Hemings, and their six children would have had all of the advantages of Jefferson's other children. Jefferson and Hemings's four children who survived into adulthood all left Monticello and merged into "white" society, marrying white partners. Sally Hemings changed her name to Sarah Jefferson when she left Monticello after Jefferson's death and lived with two of her freed sons. Is there a gene that counters the selfish gene that forces you sell your children into slavery?

A different pattern occurred in Mexico. In his old age, decades after the conquest of Mexico in 1521, Bernal Díaz del Castillo wrote *The True History of the Conquest of New Spain*. His book is one of the primary sources of Prescott's (1843) *History of the Conquest of*

Mexico and subsequent accounts. The conquistadors defeated the Aztecs through alliances with the Indian tribes that were subject to Aztec rule, taxes, and tribute. The tribute included providing victims for the sacrifices that formed part of Aztec religious practice. As the Spanish moved through villages in which young men and women were penned up in cages to be fattened for the Aztec priests who cut them open and pulled out their beating hearts, it was not surprising that the conquistadors found willing allies who fought to destroy Aztec dominance. The Greeks in the Iliad ate the animals that they had sacrificed to the Gods; the Aztecs ate the sacrifices to their gods—people.

Díaz del Castillo recorded instances in which grateful tribal leaders offered their daughters in marriage to the Spanish conquistadors. A priest performed rites that instantly converted the young women to Catholicism. The young woman Malinche who became translator, confidante, and mistress of Hernán Cortés was baptized and addressed as Dona Marina. Using the honorific "Dona," Díaz del Castillo referred to her as "the great lady." Her son was Don Martin Cortés. She married one of Cortés's lieutenants, Don Jaramillo, when Cortés returned to Spain (he had a wife in Spain), and her daughter by that marriage was known as Dona Maria. Isabel Tolosa Cortés Moctezuma, the great-granddaughter of Moctezuma the Aztec leader, became the wife of Juan de Onate—founder and governor of Nuevo Mexico.

The selfish gene seems to have been at work in Mexico, and absent in Virginia and a good part of the thirteen colonies and, until 1865, almost half of the United States of America. It often still seems to be absent. Our genetic endowment does count in shaping human behavior, but it is clear that simplistic genetic determinism cannot account for the manner in which we act and think. Nor can nonfalsifiable theories that postulate innate domain-specific language or moral organs and genes bring us to an understanding of these human qualities.

Chapter Seven

What Makes Us Tick

The previous chapters took into account a wide range of phenomena and experiments that provide insights on the nature and evolution of the brain bases of human unpredictability—the creative capacity that shapes human culture.

The Biological Bases of Unpredictability

Experiments-in-nature, studies of aphasia, showed that Broca's and Wernicke's areas are not the brain's language organ. Studies that took into account Parkinson disease and other instances of trauma to subcortical and cortical structures pointed to circuits linking different regions of the cortex and basal ganglia regulating motor control, including speech, and various aspects of cognitive acts, including language. Neural circuits linking prefrontal cortex and the basal ganglia played a key role in conferring the suite of cognitive acts subsumed under the cover-term, executive control. These include verbal and visual working memory, planning or suppressing a response, and the central act underlying creativity—being able to form and select an alternative course of action or concept.

These studies did not occur in isolation. Tracer studies of other species mapped out neural circuits. Microelectrode studies of brain activity in animals refined hypotheses concerning the local operations carried out in the subcortical basal ganglia, other subcortical structures, and cortex. The advent of noninvasive neuroimaging techniques that permit monitoring neural activity in living human subjects showed similar circuits linking prefrontal cortex and the basal ganglia in humans. Neuroimaging studies have clarified the local operations performed in these circuits. Ventrolateral prefrontal cortex, for example, is engaged in planning a cognitive shift, selecting words and memories in specific contexts, and a range of cognitive acts. Dorsolateral prefrontal cortex is engaged in monitoring or tracking cognitive acts as well as planning per se. The basal ganglia working with cortex constitute the brain's sequencing and switching engine. Comparative studies suggest that the size of the human brain has enhanced its storage capacity and perhaps its computational efficiency.

Humans do not possess unique, species-specific circuits that might account for our enhanced cognitive flexibility, the root of human unpredictability and creativity. Our knowledge of how brains work and how the human brain evolved is modest, but a serendipitous experiment-in-nature identified a gene, $FOXP2_{human}$, the human transcriptional factor that enhances associative learning and information transfer in the basal ganglia, supercharging the circuits that confer cognitive flexibility. Comparisons of the chimpanzee, human, Neanderthal, and Denisovan genomes have revealed other unique or modified genes that act on the human brain. The role of most of these genes is still unclear, but it is clearly the case that the genes that make us human are ones that contribute to cognitive flexibility and creativity, not genes that rigidly channel our thoughts and behavior.

Evolutionary Biology

There is a seeming contradiction earlier if you took the message of this book to be that genes don't count. As I pointed out on the first page, we are at the start of a new era in our understanding of how our genetic endowment shapes human behavior. The "gene game" played by practitioners of evolutionary psychology diminishes insights on how genes might influence the way that we behave and think. In contrast, as I have stressed, evolutionary biology provides the basis for advancing our understanding of what aspects of our biological endowment and their evolution may account for human uniqueness. Charles Darwin introduced the methods of evolutionary biology when he attempted to account for the "transmutation" of species in 1859. He applied these methods to the study of emotion in his 1872 book, *The Expression of the Emotions in Man and Animals,* which attempted to take into account the role of heritable, hence genetically transmitted mechanisms. Darwin sometimes went out on a limb, as when he attributed an English child's shrugging to her Parisian grandfather. The young girl had been raised in a proper English shrug-free environment! However, she shrugged, and Darwin implied that she had inherited an innate French tendency to shrug from her grandfather—in current terms, his shrug gene. This case notwithstanding, Darwin generally relied on the methods and principles of evolutionary biology. In discussing the facial expression of emotion, Darwin, for example, took note of the similar muscles of apes and humans, establishing a link to our primate heritage. We can "read" many of the expressions of animals, and as Michael Tomasello's (2009) research group has shown, dogs, apes, and other features can read some of our moods.

However, it is the case that human culture plays a part in these processes. If you have ever been engaged in conversation with someone from Nepal or India, you might have noticed their heads moving from side to side to signify agreement. As David McNeill (1985) pointed out, many of the gestures by which humans communicate both referential and emotional information are "emblematic," culturally conditioned and transmitted. One of my students got into a serious misunderstanding on a trip to India some years ago when she thought that she had forcefully signaled "no!"

Some aspects of human behavior appear to derive from "entrenched" genes. Jerry Kagan, one of the pioneers of developmental psychology, studied the biological and social bases of temperament for more than four decades. Kagan's studies suggest that shyness has a strong genetic component (Kagan, 1981). Children of shy parents tend to be shy. Children of extroverted parents are likely to be extroverted. Kagan and his colleagues in a 1988 study showed that the physiologic reactions of shy children to low levels of task-induced stress were as much as ten times greater than outgoing, extroverted children. Kagan's methods differ profoundly from the "mind-experiments" favored by some evolutionary psychologists. Kagan's research group first assessed the degree to which a child was shy by observing three-year-old children's reactions to an unusual event. Children were observed, one at a time. Each child played in a toy-filled room with her/his mother and an observer for 30 minutes or more until the child was relaxed, absorbed in play. At that point, a "spaceman" entered—a stranger dressed in a silver-colored Mylar "space-suit" with a toy helmet covering her face. The spaceman was a Harvard University graduate student. The shyest children predictably would run toward their mothers and hide behind their legs, peeking out, often in tears. In contrast, outgoing children would run toward the spaceman and sometimes even try to climb up her to remove her helmet.

A month later, the shyest and most outgoing children were monitored while they performed a simple cognitive task at increased workloads. After establishing that they regularly watched the children's TV show *Sesame Street*, they first were asked to name two characters, then four, and then answer the question, "Who chased the squirrel?" Task-induced stress manifests itself at the physiologic level in heart rate, respiratory rate, galvanic skin conduction (which reflects sweating), cortisol levels, and a speech parameter that some of my early studies showed is a measure of task-induced stress. The fundamental frequency of phonation, which determines the pitch of one's voice, has a micro-vibrato. Voice pitch always varies slightly (Lieberman, 1961). The technical term coined to describe this phenomenon is "jitter." Jitter reflects the balance of muscular tension in the larynx and correlates with stress levels. Task-induced stress decreases jitter, and we unconsciously perceive that someone is stressed by taking account of jitter (Lieberman and Michaels, 1962).

Changes in heart rate and jitter stress measures were ten times greater for the timid, shy children than for the risk-taking, extroverted children. Stephen Suomi at NIH's Poolesville Animal Center observed similar responses when shy monkeys encountered strangers. Since they were monkeys, it was possible to switch monkey infants and pair a shy infant with an outgoing monkey mother. When cross-fostered with outgoing monkey mothers, shy monkeys tended to remain shy. The reverse pattern held for outgoing monkey infants and shy monkey mothers, which again points to a genetic component to being shy or outgoing. There was a subtle effect of upbringing— monkey temperament was influenced by monkey mother's temperament; Kagan estimates about 80% heritability for shyness in humans.

What Does Genetic Variation Buy?

If shyness or being outgoing have a genetic component that dates back to monkeys, why might some humans retain genes that result in their being shy and timid, while other people have retained genes that play a part in being outgoing and willing to take risks? A misadventure that befell the Poolesville monkeys points to the answer. A chain-link fence encloses the large monkey range at NIH's primate research center. When a gap in the fence allowed monkeys to escape, timid, shy ones held back. In this instance, it paid to be shy. The adventurous extroverts who ventured out were hit by a truck as they crossed the road next to the fence. In other instances, risk-taking is appropriate. My grandparents, who left Eastern Europe at the start of the twentieth century, were not timid souls. They coped with the uncertainties of a new culture and a new language to achieve a better life for themselves and their descendants. That fortuitously turned out to be the correct choice in the "struggle for existence." They could not have foreseen the brief Nazi reign that resulted in the murder of their timid neighbors who stayed at home. In different circumstances, different modes of behavior are advantageous.

Genetic diversity, the "feedstock" for natural selection, is generally advantageous. Famines were a recurring feature of life in India

193

for thousands of years—they finally ended in the 1970s, when high-yield varieties of wheat and rice developed in the "green-revolution" that started in the 1940s were sown. However, one of the unwanted side-products of the green-revolution was the reduction of genetic diversity that occurred as native varieties of rice and wheat become extinct. Ecologists and agronomists have rightly pointed out that the loss of genetic diversity poses a danger. It isn't possible to anticipate what the future may hold in store. Genetic diversity provides a hedge. The microorganisms that are responsible for plant diseases are continually evolving, and genes that no longer exist might protect food crops from a "new" threat. Global weather conditions change, and genes that would allow crops to flourish in the hot and humid conditions that may follow from global warming will not be available if every single stalk of wheat has the same high-yield genetic profile. In consequence, seed samples are being stored in the Svalbard Global Seed Vault on the remote Norwegian Island of Spitsbergen. The underground climate-controlled cavern cost about nine million dollars to construct. The yearly operating expenses are being paid by Norway and the Global Crop Diversity Trust.

Choose Your Gene

Issues concerning human morality almost always involve aggression and altruism. Here, too, evolutionary biology suggests that simplistic solutions that link behavior to genes are misleading. Jane Goodall, who over the span of 20 years observed chimpanzee life in the Gombe Stream National Park, showed that the Gombe chimpanzees have a complex fission-fusion society. Adult males and adolescent chimpanzees often roam about in small groups throughout the territorial range, while chimpanzee mothers, their infants, and young stay put. The two groups can rejoin when a new source of food is located. In this "culture," some females have higher social ranks than others, and their social rank is passed on to their female offspring. Male chimpanzees have a more fluid social order. Males can achieve "alpha," dominant, status by displays and bites or blows that can result in severe wounds, but within the group, violence does not extend to murder.

At the start of Goodall's study, the chimpanzees formed a single group. However, warfare—organized murderous violence–broke out when part of the group split off from their fellow chimpanzees, yielding an IN and an OUT group. The members of this OUT group, formerly members of the inclusive group, established an adjoining territorial range until they were systematically hunted down and killed. A video provided by Chris Boehm, one of Goodall's colleagues, shows a chimpanzee patrol on the territorial boundary. The chimpanzees had become acclimatized to videographers who accompanied them as they both performed daily routine tasks—nursing infants, termite-fishing (fishing tools were prepared by stripping branches from a small tree limb and inserting it into a termite mound), and hunting monkeys—and went on the warpath.

You can read the intent, tense faces of the file of chimpanzees in Boehm's war video. That's what the conflict turned into—organized murderous aggression directed toward "enemies." As is the case for humans, friends can turn into enemies. The video shows a string of uncharacteristically quiet chimpanzees moving warily in single file as they approach the territorial boundary, attempting to find a lone "enemy" chimpanzee to mob and murder. In the instance videographed, the patrol met up with enemies masked by the dense forest. Neither they, nor their potential adversaries, could gauge each other's strength, so they retreated. It was calculated warfare. Humans are not alone in deciding when to fight; the chimpanzee war-party retreat suggests a cognitive risks-benefits assessment.

Wrangham and Peterson (1996) documented similar instances of chimpanzee warfare and territoriality. The chimpanzee patrols and territorial claims suggest a "primitive" genetic basis, predating the evolution of hominins, for these human attributes. However, these studies of chimpanzee life also point to the presence of cognitively mediated choices—even in chimpanzees.

Alliances and favors also can provide a path to alpha-male status, through which everyone profits. Frans de Waal's book *Chimpanzee Politics: Power and Sex among Apes*, published in 1982, described his observations of chimpanzees living in a captive colony at the Arnhem Zoo in the Netherlands. This book and its 2007 revised edition show that males can achieve alpha status by forming cooperative alliances and dispensing favors. De Waal traced the alliances formed by three male chimpanzees who rose to alpha status through politics—forming shifting coalitions that included females. These studies of chimpanzees show the different paths that can lead to alpha-male status—brute force or coalitions glued together by favors. Chimpanzee grooming is a sort of schmoozing practiced by both males and females—a chimpanzee removes debris from another's fur while patting and smoothing.

Mark Foster and his colleagues documented similar acts in the state-of-nature; their 2009 paper reviewed observations made between 1989 and 2003 of the tactics used by Gombe male chimpanzees to achieve alpha status. Large males tended to rely on violent brute force. Smaller males formed coalitions to achieve alpha status and assiduously groomed other chimpanzees, both male and female. Frodo, who weighed more than 100 pounds, bit and hit his way to power, acting the role of a despot who allowed lesser chimpanzees to groom him. Frodo almost never groomed any other chimpanzee. In contrast, Wilkie, who weighed about 80 pounds, groomed his way to alpha status. Wilkie's tactics resemble those employed by candidates to Providence's city council—obsessive hand-shaking and offers to repave the sidewalk in front of your home. No one has isolated the genes that mediate aggression or cooperation in either chimpanzees or humans, but the data discussed earlier and in other studies of primate behavior suggest genetic bases for very different courses of action. Chimpanzees in some circumstances appear to perform "cognitive" cost-benefit assessments, choosing a course of action that is most likely to achieve their goals, reflecting some degree of cognitive flexibility.

Charles Darwin may have regretted ever coining the phrase the "struggle for existence," which usually is interpreted to mean conflict. He stressed that he was using the phrase in a metaphorical sense and took pains to point out "complex relations" that enter into evolution that preclude simplistic assertions that a particular gene governs some aspect of behavior. Almost half of chapter III of *The*

Origin of Species, in which Darwin attempts to explain what the struggle for existence entails, draws on examples where

> plants and animals most remote in the scale of nature are bound together by a web of complex relations (Darwin, 1859, p. 73).

The complex relations discussed include ones holding between mistletoe and sundry flowers and birds, trees and cattle, insects and birds, orchards and moths. Darwin's message was that these interactions benefit all concerned and must entail interlocking biological factors in these different species.

Morality Briefly Revisited

The abrupt shifts in moral conduct discussed earlier in the age of the Vikings and in Germany and elsewhere during the twentieth century resulted from cognitively mediated choices shared by a culture. Humans may have inherited the genetic dispositions seen in chimpanzees toward forming IN and OUT groups. And we may also have two very different genetically mediated behavioral patterns toward achieving alpha status. However, unlike chimpanzees, we can turn on a dime, shifting the membership criteria for IN and OUT groups as well as what actions are permissible in IN or toward OUT groups.

When Japan was emerging from hundreds of years of isolation in the 1930s, anyone who was not Japanese apparently was a candidate for an OUT group. Nanking, the capital of the Republic of China, was captured by the Japanese Imperial Army on December 13, 1937. Over the next six weeks, the Imperial Army murdered hundreds of thousands of Chinese. About 80,000 men, women, and children were raped. The atrocities were witnessed by a horrified German, a member of the Nazi party, who attempted to intervene. The near-instant shift in Japanese conduct at the end of World War II clearly had no genetic basis. It perhaps can be attributed to cognitive shock-treatment resulting from a rain of incendiary bombs on

197

the wood and paper cities of Japan, the nuclear destruction of Hiro-shima and Nagasaki, and the collapse of the old social order.

The human capacity to place oneself in the place of another (the technical term is "theory of mind") is a necessary, but not a suffi-cient, condition for not doing harm. Children at first lack this con-ceptual capacity; it gradually develops, as is the case for other aspects of cognition. A primitive theory of mind is often equated with the "mirror-test," in which a chimpanzee can recognize that she is view-ing herself in a mirror, but apes act in a manner that they themselves would find unacceptable. The Gombe chimpanzee video records include a hunting party sharing part of a kill. A screaming live mon-key is the tidbit being passed from ape to ape as they each take a bite. Humans often act in worse manner.

Our human ability to place oneself in the place of another can enhance brutality, inflicting degrading torture that no ape can con-ceive of, as well as acts in accord with the "Golden Rule." It took a long time for the Golden Rule to be codified. In the Hebrew Bible, which was written over thousands of years, the code of Jewish con-duct changed from the celebration of King David's wars of exter-mination (whether they actually occurred or not) to the teachings of Hillel, who lived at about the same time as the historical Jesus. Hillel summed up the meaning of Torah, the Hebrew bible, as

> that which is hateful to you, do not do to your fellow. That
> is the whole Torah; the rest is the explanation; go and learn.

The same injunction, rephrased in the gospel of Matthew, is the usual form of the Golden Rule:

> Do unto others as you would have them do unto you.

It took a long time for "enlightened" WEIRDO cultures to act on the Golden Rule. In 1745, after the battle of Culloden, three Scottish rebels were first hanged by the neck at the Tower of Lon-don. After 14 minutes, they were cut down, still alive, and disem-boweled. Their bodies were then quartered. Thirty-one years later, King George's government, anticipating the execution of Benjamin Franklin, Thomas Jefferson, George Washington, and their fellow rebels modified the schedule of punishment, eliminating disembowelment. However, whether this change reflected a change in moral attitude, or efficiency, it was never carried out, because the rebels morphed into the statesmen of the United States of America. The Golden Rule still doesn't seem to characterize human conduct.

Is It a Boy or Girl on the Phone?

Something as disarmingly simple as deciding whether it is a small boy or girl speaking to you on the telephone illustrates the intersection of genetic endowment and culture. Girls' names usually signal that they are a girl, but that isn't always the case. But when you hear a girl speaking on the telephone, you usually can tell that a girl talking to you, apart from what she might be talking about. That can be the case, even when she is five years old. However, when one of our youngest son's friends phoned, I was certain when I answered the phone that it was a girl, which turned out to be wrong. The acoustic signal somehow had indicated that it was a girl. I was then working with Edmund S. Crelin at Yale University, whose specialty was human anatomical development. When I mentioned this odd misidentification in passing to Ed, he, too, was puzzled, because there are no anatomical differences at age five that would cause a girl's voice to have acoustic characteristics that are different from a boy's. Why did I think that a young girl was on the phone?

Jacquie Sachs, who was a member of the faculty of the University of Connecticut's Department of Speech; Donna Erikson, one of my graduate students; and I decided to study a group of children and acoustically analyze their speech. My boy/girl error suggested that the acoustic distinctions might reflect gender rather than biological sex, and that supposition was confirmed. We recorded 30 five-year-old boys and girls. The children were recorded as they chattered about what they were going to do that day and as they named pictures of animals. When short segments of their speech were presented to a panel of adult listeners, the boys were all identified as boys, but three girls were identified as boys. The girls were said to be tomboys by their parents. My son's friend wasn't among our subjects.

Acoustic analysis showed that there was no systematic difference between the boys' and girls' fundamental frequencies of phonation—the acoustic determinant of pitch. As chapter 2 pointed out, the pitch of a person's voice is our perceptual response to the average fundamental frequency of phonation (F_0), the rate at which the vocal cords of the larynx open and close. The similar range of voice pitch for the boys and girls wasn't surprising. It is common knowledge that boys' voices don't change until puberty, when the male larynx becomes larger, yielding lower-pitched voices. What then signaled that it was a boy or girl talking, or rather a child who wanted to be perceived as a boy or girl? Acoustic analysis revealed different average formant frequencies for "boy" or "girl" voices. Boy vowels had lower formant frequencies than the same girl vowels. This difference also was unexpected because at age five there didn't seem to be a systematic difference in the length of boys' and girls' supralaryngeal vocal tracts (SVTs)—the distance between their lips and larynx (since confirmed by D. Lieberman and McCarthy, 1999, and D. Lieberman et al., 2001). The length of the SVT, as you saw in chapter 2, determines the absolute value of vowel formant frequencies. Longer SVTs produce lower formant frequencies.

Observations of the children's lips as they talked solved the mystery. Boys produced a male vowel "dialect" by slightly "rounding" their lips as they talked, protruding and closing them ever so slightly. These lip maneuvers lowered their vowel formant frequencies. In contrast, the girls' female vowel dialect resulted from children retracting their lips, smiling ever so slightly, which raised their formant frequencies. In effect, boys modify their vowels to "sound" bigger, girls "smaller." Our son's friend on the telephone, who had three sisters, apparently had picked up the wrong dialect. Subsequent studies have replicated these gender-specific gestures for children living in the United States. It is not clear whether this is a "universal" aspect of human behavior because systematic studies of children living in other cultures have yet to be done. Formant frequencies provide a better cue than pitch for conveying gender because there is a great deal of overlap between the pitch ranges of men and women. Moreover, it is almost impossible to systematically raise or lower your pitch during normal conversations.

The perceptual difference between the American-English male and female speech dialects might best be described in musical terms as timbre. The dialects serve to signal gender. Why did the children act in this manner? Human conduct suggests that it is one of the many ways that biology and culture intersect. Male and female birds generally have different plumage. Size and morphology differentiates male from female chimpanzees, gorillas, and orangutans, but humans as a species have less secondary sexual dimorphism than apes. In very different human cultures, clothing and manner instead serve to signal gender, usually, but not always coincident with biological sex. In societies where males and females have specified tasks that are necessary for survival, such as in traditional Greenland Inuit society, girls will affect male clothes, act like boys, and carry out a male-specified task such as hunting seals in families that have daughters but no sons, until they are about to marry. An anthropologist who lived with an Inuit family once related how she was startled when she was invited to the marriage of the girl she had thought was a boy up to a week before the event.

Boy versus girl voice qualities reflect a "primitive" biological, genetic reality. Adult males as a group have longer vocal tracts and are bigger than adult females. The five-year-old children were modeling their behavior on these archetypes, but they didn't have to—no one was instructing them to do so, and there was an element of choice involved. The "tomboys" didn't adopt the female register.

I used the term "primitive" in its evolutionary sense—a characteristic that a species shares with ancestral species. In this instance our voice is signaling how big we are, in a manner similar to the actions of monkeys, deer, and other species. Almost 30 years elapsed before this possibility suggested itself. A series of studies by W. Tecumseh Fitch, before he became preoccupied with Universal Grammar and recursion, shows that many animals estimate how big a conspecific is, by taking account of the absolute value of formant frequencies (Fitch, 2010). Male deer pull their larynx down as they vocalize to make themselves sound bigger to other deer, (Fitch, 2000). McElligott et al. (2006) writing in the *Journal of Zoology*, describe male deer calls during mating season as "groans" that are pleasing to does. I don't know of any human courtship ritual that involves men groaning to attract a mate, but the formant-frequency gender specifying effects that marked five-year-olds in Storrs, Connecticut, in 1972 and other young children (mostly WEIRDO kids, owing to where our study has been replicated) seem to have antecedents in the behavior of other species.

Culture as an Agent of Genetic Change

It should be clear that it is inherently impossible to completely partition the contributions of culture and genes to behavior. Culture is the agent that shapes human ecosystems and hence intersects with biology at the genetic and epigenetic level. Epigenetics, the study of instances in which the DNA sequences that constitute the genetic code do not change but the expression of genes changes, also points to cultural-genetic interaction. Studies of mammalian species running the gamut from rodents to humans show that environmental factors can have epigenetic effects, changing the expression of genes. Rat mothers, for example, clean and groom their pups after birth, and some rats groom their pups to a much greater extent than other rat mothers. The mothers' licking turns out to trigger the pup's hippocampus, which, in turn, releases a sequence of hormones that change the expression of genes. Pups that are groomed to a greater extent are healthier, grow faster, and are "better" rats than those born to rat mothers who groom their pups to a lesser extent (Zhang et al., 2010). Marcus Pembry, who was a member of the research team that studied the KE family and led to the discovery of the FOXP2 transcriptional factor, is the lead author of a study that points to epigenetic effects in humans. Pembry and his colleagues studied the grandsons of Swedish men who experienced famines as children during the closing years of the nineteenth century. Sweden then was a poor country that had little in the way of a social safety net, accounting for the thousands of Swedes who migrated to the United States. The grandsons of Swedish men who grew up during these nineteenth-century famines are less likely to die of cardiovascular diseases, but are more susceptible to diabetes. The granddaughters of Swedish women whose mothers experienced famine while bearing them lived shorter lives. Behavioral data also suggest epigenetic effects on intellectual development. Jerry Kagan found that impoverished, narrow environmental stimulation during the first year of life impeded a child's intellectual development, but the delays could be removed by a richer, stimulating, environment during the second year of life (Kagan, 1981).

Current research surprisingly shows that the "social" environment can even have a direct effect at the genetic level, such as the release of transcriptional factor Foxp2. Zebra finches, as noted earlier, are the birds of choice for research on the behavioral effects of the Foxp2 transcriptional factor. Zebra finches learn new songs each year, and the avian form of Foxp2 peaks in the birds' circuit to area X of the bird brain (Brainard and Doupe, 2000; Doupe et al., 2005). This circuit is homologous with the cortical-basal ganglia neural circuits implicated in associative learning in mammals. Surprisingly, Foxp2 is at a higher level when a male bird is actually singing to a potential mate, than when the he is alone, practicing his songs (Teramatsu and White, 2006).

Culture, Language, and Thought

Finally, though biology and hence genetics determines baseline human cognitive capacities, there are intimate, complex relationships that hold between biology, culture, language, and thought. Biology sets limits on thought, but culture changes biology, language transmits culture, and culture influences language and thought. Paradoxically, current research that is unraveling these processes started with the supposition that language constrains thought.

In the middle years of the twentieth century, the American linguists Edward Sapir and Benjamin Whorf claimed that language shaped and fixed your worldview. In the extreme version of their theory, "Linguistic Relativity," a person's language placed absolute limits on his or her conceptual framework. The roots of the Sapir-Whorf hypothesis can be traced back to the early years of the nineteenth century. Wilhelm von Humboldt in 1835 took an even more extreme view of the effects of language on thought. Humboldt, in his opus *On Language: The Diversity of Human Language-Structure and Its Influence on the Mental Development of Mankind*, set forth the premise that language determines the intellectual attainments of a culture. To Humboldt, language and cognition were indivisible:

> Language is the formative organ of *thought*. . . . Thought and language are therefore one and inseparable from one another (Humboldt [1835] 1988, p. 46).

Humboldt had some peculiar views on how language affects thought. Humboldt was taken with the notion that inflected languages, such as Latin and German, that modify a word to indicate whether a person or thing is the subject or object of a verb, are superior instruments for thought than uninflected languages that instead use word order for that end. In Humboldt's day, Sanskrit was thought to be the end point of inflected languages. Sanskrit to Humboldt therefore marked minds capable of the greatest intellectual achievements. German, Humboldt's native language, was pretty close. In contrast, Humboldt asserted that Chinese, at the other end of the inflected-uninflected scale, which relies on word order to convey who is doing something to whom or what, was an inferior vehicle of thought. Chinese-speakers to Humboldt therefore had inferior cognitive abilities—a claim that reflects Wilhelm von Humboldt's profound ignorance of the achievements of Chinese civilization. Wilhelm von Humboldt, unlike his brother, Alexander von Humboldt, the early nineteenth-century explorer and naturalist, never lived or traveled outside of Western Europe.

The Sapir-Whorf hypothesis has been hotly disputed. Many of the phenomena cited to support the Sapir-Whorf hypothesis did not hold when reexamined. Eskimos (Inuits) were supposed to be able to perceive finer distinctions between different types of snow than monolingual speakers of English because the Inuit language had many words that described different types of snow, but that wasn't the case (Pullum, 1991). Whether you have many words to describe snow conditions or a few doesn't really affect the fact that virtually anyone can perceive that the snow on the ground is soft or hard. However, having words that convey these distinctions does make it simpler to communicate snow conditions. English-speaking cross-country skiers have to take into account snow conditions and therefore coined and used "new" compound words such as "blue-wax-snow," "green-wax-snow," "red-wax-snow" to communicate distinctions in snow that are of interest to skiers. A culture's needs are reflected in language. Linguistic relativity was declared dead after study after study refuted Whorf's claims, but the focus of these studies was simplistic. It isn't simply the case that thought is constrained by language; language reflects the needs of a culture and also affects the way that the members of that culture view the world.

Language as a Tool

Daniel Everett's 2005 report on the language and culture of the Piraha people, who live in near isolation in Brazil's vast Amazonian region, was a bombshell in the hermetic world of linguistic scholarship. The process of recursion, which to Noam Chomsky, Marc Hauser, Tecumseh Fitch, and other like-minded linguists was supposed to be the feature that singles out language from all other aspects of behavior, doesn't seem to exist in the Piraha's language. The Piraha don't have a written language, which isn't strange when you reflect on the fact that until recently, say 10,000 years ago, that was the case for everyone. But even illiterate people when they speak produce sentences, and Piraha sentences never include relative clauses, embedded clauses of any kind, or even the conjoined sentences that four-year-old children speaking other languages produce.

Everett showed that Piraha uses only simple sentences; recursion does not exist in the Piraha language. The reaction to Everett's 2005 *Current Anthropology* paper and his 2008 book ranged from some dedicated adherents of Noam Chomsky calling Everett a liar, which is not usual in scholarly discourse, to Chomsky's (2012a, p. 1) finessing the problem by proposing that the "range" R of recursion could be "null," thereby sweeping the proposition that recursion is the defining property of human language under the carpet.

But the focus of Everett's work is not on Noam Chomsky or recursion. Everett's 2012 book *Language: The Cultural Tool* makes it clear that he views language as a tool that is crafted to meet the needs and values of a culture. The Piraha comfortably live in a very relaxed manner on the bounty of the river and fertile land. Day, night, moonlight, rainfall, and the level of water of the river set the rhythm of life. As Everett, who lived with the Piraha for a total of 86 months, points out:

> The Piraha tend to talk almost exclusively about the here and now. They do not celebrate birthdays or anniversaries. They do however, recognize the passage of time through wet and dry seasons, and use the full moon as a simple calendar.

In consequence, they have 12 words that refer to time: *ahoapio'* signifies another day; *ahoa'i* night; *a'hoakohosihio*, early morning; *pi'i*, now; and words for during the day, noon, sunset, full moon. A month is roughly "moon." There is no word for a week. A year is "water," referring to a cycle of high and low water in the river. Two words, *pila'iso* and *piibigaiso*, signify low water or high water. Beyond that, the Piraha don't seem to concern themselves with time, and their language doesn't have tense markers that convey present and past actions.

The Piraha language also seems odd to me and probably to you in that it has no explicit color terms or numbers. The color of something is simply described as the color of X, where X is another item in immediate view or in shared knowledge. Everett attempted to teach Piraha young and old to count—he failed. In light of the fact that pigeons can be taught to count up to nine items (Scarf et al., 2011), it is unlikely that any mental deficiency is responsible for the Piraha being arithmetically challenged. Nor is it likely that they lack the "number faculty" proposed by Chomsky (1988, p. 168). According to Chomsky, children do not learn to add numbers; innate knowledge of arithmetic is also preloaded into our brains. However, the Piraha language meets their needs. There is no genetic basis for these features of the Piraha language. Children abducted by other tribes have no difficulty learning other languages, including Brazilian Portuguese. Piraha gene frequencies, moreover, include those of other groups, owing to their custom of offering daughters or wives to visiting males.

The Piraha language reflects the needs of their culture and constrains their view of life. The absence of past, present, and future tense markers in their language reflects their living in and being concerned with the present. Everett initially came into the world of the Piraha as a missionary to bring the word of Jesus Christ to them. He learned their language so that he could translate and bring the good news of the Bible to them. The conceptual constraints of Piraha culture, manifested in their language, were evident when he completed his translation of the Gospel of John. The reaction of the Piraha when he read it aloud was to ask where and when he had met John. On being told that he had never met John, they had no further interest in the teachings of John. Everett had not personally heard John telling his story, and the story therefore could not be true. The Piraha have no sense or record of history beyond the lifespan of a living observer; events that have not been witnessed by a living person have no standing. Piraha bards reciting tales of the distant past do not exist.

Language to Everett is the mirror of culture; the concepts coded in a language are tools that play a useful role in a culture. Everett's books are delightfully jargon-free, so you can best appreciate the complexities that hold between culture, language, and thought by reading them.

The Genes That Count

The field of psychology tends to continuously fall for comprehensive theories—Freud, Skinner, Jung—that don't pan out empirically. That also is the case for Chomskian linguistics and the school of evolutionary psychology that relies so heavily upon it. We supposedly have an innate moral organ transmitted by a gene or genes that shape a "Universal Moral Grammar" similar to the Universal Grammar instantiated in Chomsky's Faculty of Language. Evolutionary psychology proposes similar answers for other human attributes. Art, a human quality whose qualities are perhaps even more elusive, supposedly results from another module that yields our "art instinct."

When these theories are not patently absurd, such as an art-instinct theory that predicts that everyone on earth prefers images of a grassy plain with a few trees and water to all other scenes, they are just-so stories, crafted so as to be impossible to falsify and hence predict nothing. Instead, virtually everything that characterizes the way that we live is the product of human culture, transmitted from one generation to the next through the medium of language. But the form of one's language is culturally transmitted, again through the medium of language. And, in turn, culture shapes the human ecosystem and can lead to genetic evolution.

We are at the starting point, not the end, of the quest to understand how the human brain works and was shaped. However, what is clear is that we are not ruled by genes that were fixed in earlier stages of human evolution, or genes inherited from our primate ancestors, or invented genes that never were. It is becoming apparent that the genes setting us apart from our primate cousins act to enhance our cognitive capabilities and confer our cognitive flexibility—the engine of human creativity that allows us to shape our destiny. We possess the ability to change the manner in which we act toward each other and how we view the world around us. It is precisely because we are so unpredictable that humans are unique.

We are the unpredictable species.

References

Abarbanell, L., and M. D. Hauser. 2010. Mayan morality: an exploration of permissible harms. *Cognition* 115: 207–224.

Abi-Rached, L., M. J. Jobin, S. Kulkarni, A. McWhinnie, K. Dalva , L. Gragert, et al. 2011. The shaping of modern human immune systems by multiregional admixture with archaic humans. *Science* 334: 89–98.

Aldridge, W., and K. C. Berridge. 1998. Coding of serial order by neostriatal neurons: a "natural action" approach to movement sequence. *The Journal of Neuroscience* 18: 2777–2787.

Alexander, G. E., M. R. DeLong, and P. L. Strick. 1986. Parallel organization of segregated circuits linking basal ganglia and cortex. *Annual Review of Neuroscience* 9: 357–381.

Altmann, J. 1980. *Baboon Mothers and Infants.* Cambridge, MA: Harvard University Press.

Altmann, S., and J. Altmann. 1971. *Baboon Ecology: African Field Research.* Chicago: University of Chicago Press.

Amunts, K., A. Schleicher, U. Burgel, H. Mohlberg, H.B.M. Uylings, and K. Ziles. 1999. Broca's area revisited: cytoarchitecture and intersubject variability. *The Journal of Comparative Neurology* 412: 319–341.

Arnett, J. 2008. Why American psychology needs to become less American. *American Psychologist* 63: 602–614.

Assad, W. F., and E. N. Eskander. 2011. Encoding of both positive and negative reward prediction errors by neurons of the primate lateral prefrontal cortex and caudate nucleus. *The Journal of Neuroscience* 31: 17772–17787.

Bacon, F. 1620. Internet rendition based on the standard translation of James Spedding, Robert Leslie Ellis, and Douglas Denon Heath in *The Works*, vol. VIII, published in Boston by Taggard and Thompson in 1863.

Baddeley, A. D. 1986. *Working Memory*. Oxford: Clarendon.

Badre, D., and A. D. Wagner. 2006. Computational and neurobiological mechanisms underlying cognitive flexibility. *PNAS* 103: 7186–7190.

Barbujani, G., and R. R. Sokal. 1990. Zones of sharp genetic change in Europe are also linguistic boundaries. *Proceedings of the National Academy of Sciences, USA* 187: 1816–1819.

Bar-Gad, I., and H. Bergman. 2001. Stepping out of the box: information processing in the neural networks of the basal ganglia. *Current Opinion in Neurobiology* 11: 689–695.

Bear, M. F., L. N. Cooper, and F. F. Ebner. 1987. A physiological basis for a theory of synaptic modification. *Science* 237: 42–48.

Beckman, M. E., T.-P. Jung, S.-H. Lee, K. de Jong, A. K. Krishnamurthy, S. C. Ahalt, et al. 1995. Variability in the production of quantal vowels revisited. *Journal of the Acoustical Society of America* 97: 471–489.

Belle, E.M.S., and G. Barbujani. 2007. A worldwide analysis of multiple microsatellites suggests that language diversity has a detectable influence on DNA diversity. *American Journal of Physical Anthropology* 133: 1137–1146.

Blair, R.J.R. 2008. The amygdale and ventrolateral prefrontal cortex: functional contributions and dysfunction in psychopathology. *Philosophical Transactions Royal Society of London B Biological Sciences* 363: 2557–2565.

Blumstein, S. E., W. E. Cooper, H. Goodglass, S. Statlender, and J. Gottlieb. 1990. Production deficits in aphasia: a voice-onset time analysis. *Brain and Language* 9: 153–170.

Boesch, C. 1993. Aspects of transmission of tool-use in wild chimpanzees. In *Tools, Language and Cognition in Human Evolution*, ed. R. Gibson and T. Ingold. New York: Cambridge University Press, 171–184.

Bouhuys, A. 1974. *Breathing*. New York: Grune and Stratton.

Bowerman, M. 1990. Talk given at the Max Planck Institute for Psycholinguistics, Nijmegen, The Netherlands.

Brainard, M. S., and A. J. Doupe. 2000. Interruption of a basal-ganglia-forebrain circuit prevents plasticity of learned vocalizations. *Nature* 404: 762–766.

Broca, P. 1861. Nouvelle observation d'aphémie produite par une lésion de la moitié postérieure des deuxième et troisième circonvolutions frontales. *Bulletin Societé Anatomie*, 2nd ser., 6: 398–407.

Brodmann, K. 1908. Beiträge zur histologischen Lokalisation der Grosshirnrinde. VII. Mitteilung: Die cytoarchitektonische Cortexgleiderung der Halbaffen (Lemuriden). *Journal für Psychologie und Neurologie* 10: 287–334.

———. 1909. Vergleichende Lokalisationslehre der Groshirnrinde in iheren Prinzipien dargestellt auf Grund des Zellenbaues. Leipzig: Barth.

———. 1912. Ergebnisse uber die vergleichende histologische Lokalisation der Grosshirnrinde mit besonderer Berucksichtigung des Stirnhirns. *Anatomischer Anzeiger*, suppl., 41: 157–216.

Bronowski, J. 1998. *The Origins of Knowledge and Imagination*. New Haven, CT: Yale University Press.

Browne, J. 1995. *Charles Darwin Voyaging*. Princeton, NJ: Princeton University Press.

Buckner, R. L. 2010. Human functional connectivity: new tools, unresolved questions. *Proceedings of the National Academy of Sciences, USA* 107(24): 10769–70.

Burling, R. 2002. The slow growth of language in children. In *The Transition to Language*, ed. A. Wray. Oxford: Oxford University Press, 297–311.

Burton, H. 2003. Visual cortex activity in early and late blind people. *The Journal of Neuroscience* 23: 4005–4011.

Calvin, W. 2002. *A Brain for All Seasons: Human Evolution and Abrupt Climate Change*. Chicago: University of Chicago Press.

Carew, T. J., E. T. Waters, and E. R. Kandel. 1981. Associative learning in aplysia: cellular correlates supporting a conditioned fear hypothesis. *Science* 211: 501–503.

Carre, R., B. Lindblom, and P. MacNeilage 1995. Acoustic factors in the evolution of the human vocal tract. *C. R. Academie des Sciences Paris*, ser. IIb, t320: 471–476.

Chapais, B. 2008. *Primeval Kinship*. Cambridge, MA: Harvard University Press.

Chie, U., I. Yuichi, M. Kimura, E. Kirino, S. Nagaoka, M. Abe, T. Nagata, and H. Arai. 2002. Irreversible subcortical dementia following high altitude illness. *High Altitude Medicine and Biology* 5: 77–81.

The Chimpanzee Sequencing and Analysis Consortium. 2005. Initial sequence of the chimpanzee genome and comparison with the human genome. *Nature* 437: 69–87.

Chomsky, C. 1969. *The Acquisition of Syntax in Children from 5 to 10*. Cambridge, MA: MIT Press.

Chomsky, N. 1957. *Syntactic Structures*. The Hague: Mouton.

———. 1972. *Language and Mind*. San Diego, CA: Harcourt Brace Jovanovich.

———. 1975. *Reflections on Language*. New York: Pantheon.

———. 1976. On the nature of language. In *Origins and Evolution of Language and Speech*, ed. H. B. Steklis, S. R. Harnad, and J. Lancaster. New York: New York Academy of Sciences, 46–57.

———. 1980a. Initial states and steady states. In *Language and Learning: The Debate between Jean Piaget and Noam Chomsky*, ed. M. Piattelli-Palmarini. Cambridge, MA: Harvard University Press, 107–130.

———. 1980b. Rules and representations. *Behavioral and Brain Sciences* 3: 1–61.

———. 1986. *Knowledge of Language: Its Nature, Origin, and Use*. New York: Prager.

———. 1988. *Language and Problems of Knowledge: The Managua Lectures*. Cambridge, MA: MIT Press.

———. 1995. *The Minimalist Program*. Cambridge, MA: MIT Press.

———. 2000. *New Horizons in the Study of Language and Mind*. New York: Cambridge University Press.

———. 2012a. *Minimal Recursion: Exploring the Prospects*. Talk at the University of Massachusetts, Amherst (scheduled for publication).

———. 2012b. *The Science of Language (Interviews with James McGilvray)*. Cambridge, UK: Cambridge University Press.

Cools, R., R. A. Barker, G. J. Sahakian, and T. W. Robbins. 2001. Mechanisms of cognitive set flexibility in Parkinson's disease. *Brain* 124: 2503–2512.

Coop, G., K. Bullaughey, F. Luca, and M. Przeworski. 2008. The timing of selection at the human FOXP2 gene. *Molecular Biology and Evolution* 25: 1257–1259.

Cosmides, L., and J. Toobey. 1992. *The Adapted Mind: Evolutionary Psychology and the Generation of Culture*. New York: Oxford University Press.

Crelin, E. S. 1969. *Anatomy of the Newborn:An Atlas*. Philadelphia: Lea and Febiger.

Cummings, J. L. 1993. Frontal-subcortical circuits and human behavior. *Archives of Neurology* 50: 873–880.

Curtiss, S. 1977. *Genie: A Psycholinguistic Study of a Modern-Day "Wild Child."* New York: Academic Press.

Darwin, C. [1859] 1964. *On the Origin of Species*. Cambridge, MA: Harvard University Press.

———. 1872. *The Expression of the Emotions in Man and Animals*. London: John Murray.

Dawkins, R. 1976. *The Selfish Gene*. New York: Oxford University Press.

Deacon, T. W. 1997. *The Symbolic Species: The Co-Evolution of Language and the Human Brain*. New York: W. W. Norton.

De Boer, B. 2010. Modeling vocal anatomy's significant effect on speech. *Journal of Evolutionary Psychology* 8: 351–366.

De Boer, B., and W. T. Fitch. 2010. Computer models of vocal tract evolution: an overview and critique. *Adaptive Behavior* 18(1): 36–47.

De Waal, F. 2007. *Chimpanzee Politics: Power and Sex among Apes*, rev. ed. Baltimore: Johns Hopkins University Press.

D'Esposito, M., and M. P. Alexander. 1995. Subcortical aphasia: distinct profiles following left putaminal hemorrhage. *Neurology* 45: 38–41.

Devlin, J. T., and R. A. Poldark. 2007. In praise of tedious anatomy. *Neuroimage* 37: 1033–1058.

Dietrich, G. L., and C. W. Hinnicutt. 1948. Art content preferred by primary-grade children. *The Elementary School Journal* 48: 55–62.

Dobzhansky, T. 1973. Nothing in biology makes sense except in the light of evolution. *American Biology Teacher* 35: 125–129.

Doupe, A. J., D. J. Perkel, A. Reiner, and E. A. Stern. 2005. Birdbrains could teach basal ganglia research a new song. *Trends in Neuroscience* 28: 353–363.

Dronkers, N. F., O. Plaisant, M. T. Iba-Zizain, and E. A. Cananis. 2007. Paul Broca's historic cases: high resolution MR imaging of the brains of Leborgne and Lelong. *Brain* 130: 1432–1441.

Dunbar, R.I.M. 1988. *Primate Social Systems*. London: Chapman and Hall.

Duncan, J., and A. M. Owen. 2000. Common regions of the human frontal lobe recruited by diverse cognitive demands. *TINS* 10: 475–483.

Dutton, D. 2009. *The Art Instinct: Beauty, Pleasure and Human Evolution*. London: Bloomsbury Press.

Enard, W., M. Przeworski, S. E. Fisher, C.S.L. Lai, V. Wiebe, T. Kitano, A. P. Monaco, and S. Paabo. 2002. Molecular evolution of FOXP2, a gene involved in speech and language. *Nature* 41: 869–872.

Enard, W., S. Gehre, K. Hammerschmidt, S. M. Hölter. T. Blass, M. Somel, et al. 2009. A humanized version of FOXP2 affects cortico-basal ganglia circuits in mice. *Cell* 137: 961–971.

Evans, N., and S. C. Levinson. 2009. The myth of language universals: language diversity and its importance for cognitive science. *Behavioral and Brain Sciences* 32: 429–448.

Everett, D. L. 2005. Cultural constraints on grammar and cognition in Piraha: another look at the design features of human language. *Current Anthropology* 46: 621–634.

———. 2008. *Don't Sleep, There Are Snakes: Life and Language in the Amazonian Jungle.* London: Profile Books.

———. 2012. *Language: The Cultural Tool.* New York: Pantheon.

Falk, J. H., and J. D. Balling. 2010. Evolutionary influences on human landscape preference. *Environment and Behavior* 42: 479–493.

Fant, G. 1960. *Acoustic Theory of Speech Production.* The Hague: Mouton.

Fedorenko, E., M. K. Behr, and N. Kanwisher. 2011. Functional specificity for high-level linguistic processing in the human brain. *PNAS* 108, no. 39: 16428–16433.

Feinberg, M. J., and O. Ekberg, 1990. Deglutition after near-fatal choking episode: radiologic evaluation. *Radiology* 176: 637–640.

Finlay, B., and R. Darlington. 1995. Linked regularities in the development and evolution of mammalian brains. *Science* 268: 1578–1584.

Fisher, S. E., F. Vargha-Khadem, K. E.Watkins, A. P. Monaco, and M. E. Pembrey. 1998. Localization of a gene implicated in a severe speech and language disorder. *Nature Genetics* 18: 168–170.

Fitch, W. T. 2000. Skull dimensions in relation to body size in nonhuman mammals: the causal bases for acoustic allometry. *Zoology*,103: 40–58.

———. 2010. *The Evolution of Language.* New York: Cambridge University Press.

Fitch, W. T., and M. D. Hauser. 2004. Computational constraints on syntactic processing in a nonhuman primate. *Science* 303: 377–380.

Fitch, W. T., M. D. Hauser, and N. Chomsky. 2005. The evolution of the language faculty: clarifications and implications. *Cognition* 97(2): 179–210.

Flowers, K. A., and C. Robertson. 1985. The effects of Parkinson's disease on the ability to maintain a mental set. *Journal of Neurology, Neurosurgery, and Psychiatry* 48: 517–529.

Fodor, J. A. 1983. *Modularity of Mind.* Cambridge, MA: MIT Press.

———. 1988. *Concepts: Where Cognitive Science Went Wrong.* New York: Oxford University Press.

Fodor, J., and M. Piatelli-Palmarini. *What Darwin Got Wrong.* New York: Farrar Straus and Giroux.

Fogel, R. W. 2004. *The Escape from Hunger and Premature Death 1700–2100—Europe, America and the Third World.* New York: Cambridge University Press.

Foster, M. W., I. C. Gilby, C. M. Murray, A. Johnson, E. E. Wroblewski, and A. A. Pusey. 2009. Alpha male chimpanzee grooming patterns: implications for dominance "style." *American Journal of Primatology* 71: 136–144.

François-Brosseau, F.-E., K. Martinu, A. P. Strafella, M. Petrides, F. Simard, and O. Monchi. 2009. Basal ganglia and frontal involvement in bilateral self-generated and externally-triggered finger movements. *European Journal of Neuroscience* 29: 1277–1286.

Frederici, A. D. 2002. Towards a neural basis of auditory sentence processing. *Trends in Cognitive Science* 6: 78–84.

Gardner, R. A., and B. T. Gardner. 1969. Teaching sign language to a chimpanzee. *Science* 165: 664–672.

Gentner T. Q., K. M. Fenn, D. Margoliash, and H. C. Nubaum. 2006. Recursive syntactic pattern learning by songbirds. *Nature* 440: 1204–1207.

Geschwind, N. 1970. The organization of language and the brain. *Science* 170: 940–944.

Gibbons, A. 2012. Ancient DNA: A crystal-clear view of an extinct girl's genome. *Science* 337: 1028–1029.

Gold, E. M. 1967. Language identification in the limit. *Information and Control* 10: 447–474.

Goldstein, K. 1948. *Language and Language Disturbances.* New York: Grune and Stratton.

Goodall, J. 1986. *The Chimpanzees of Gombe: Patterns of Behavior.* Cambridge, MA: Harvard University Press.

Gopnik, M. 1990. Dysphasia in an extended family. *Nature* 344: 715.

Gottlieb, G. 1975. Development of species identification in ducklings: nature of perceptual deficits caused by embryonic auditory deprivation. *Journal of Comparative and Physiological Psychology* 89: 387–389.

Gould, S. J. 1985. Not necessarily a wing. *Natural History* 94 (October): 12–25.

Gould, S. J., and R. C. Lewontin. 1979. The spandrels of San Marco and the Panglossian program: a critique of the adaptationist programme. *Proceedings of the Royal Society of London* 205: 281–288.

Graybiel, A. M. 1995. Building action repertoires: memory and learning functions of the basal ganglia. *Current Opinion in Neurobiology* 5: 733–741.

———. 1997. The basal ganglia and cognitive pattern generators. *Schizophrenia Bulletin* 23: 459–469.

Green, R. E., J. Krause, A. W. Briggs, T. Maricic, U. Stenzel, M. Kircher, et al. 2010. A draft sequence of the Neanderthal genome. *Science*, 328: 710–722.

Greenberg, J. 1963. *Universals of Language*. Cambridge, MA: MIT Press.

Greene, J. D., F. A. Cushman, L. E. Stewart, K. Lowenberg, L. E. Nystrom, and J. D. Cohen, 2009. *Cognition* 111: 364–371.

Gross, M. 1979. On the failure of generative grammar. *Language* 55: 859–885.

Grossman, M., S. Carvell, M. B. Stern, S. Gollomp, and H. I. Hurtig. 1992. Sentence comprehension in Parkinson's disease: the role of attention and memory. *Brain and Language* 42: 347–384.

Grossman, M., S. Carvell, S. Gollomp, M. B. Stern, M. Reivich, D. Morrison, A. Alavi, and H. I. Hurtig. 1993. Cognitive and physiological substrates of impaired sentence processing in Parkinson's disease. *Journal of Cognitive Neuroscience* 5: 480–498.

Grossman, M., J. Glosser, J. Kalmanson, M. B. Morris, H. Stren, and H. I. Hurtig. 2001. Dopamine supports sentence comprehension in Parkinson's disease. *Journal of the Neurological Sciences* 184: 123–130.

Hamann, S., R. A. Herman, C. L. Nolan, and K. Wallen. 2004. Men and women differ in amygdala response to visual sexual stimuli. *Nature Neuroscience* 7: 411–416.

Harrington, D. L., and K. Y. Haaland. 1991. Sequencing in Parkinson's disease: abnormalities programming and controlling movement. *Brain* 114: 99–115.

Harris, S., J. T. Kaplan, A. Curiel, S. Y. Bookheimer, M. Iacoboni, and M. S. Cohen. 2009. The neural correlates of religious and noreligious belief. *PLoS ONE* doi: 10.1371/journal.pone.0007272.

Hauser, M. D. 2006. *Moral Minds: How Nature Designed a Universal Sense of Right and Wrong* New York: Harper Collins/Ecco.

Hauser, M.D., N. Chomsky, and W. T. Fitch. 2002. The faculty of language: what is it, who has it, how did it evolve? *Science* 298: 1569–1579.

Hazy, T. E., M. J. Frank, and R. C. O'Reilly. 2006. Banishing the homunculus: making working memory work. *Neuroscience* 139: 105–118.

Hebb, D. O. 1949. *The Organization of Behavior: A Neurobiological Study*. New York: Wiley.

Herculano-Houzel, S. 2009. The human brain in numbers: a linearly scaled-up primate brain. *Frontiers in Human Neuroscience* 3: 1–11.

Heim, J.-L. (1989). La nouvelle reconstitution du crane neanderthalien de la Chapelle-aux-Saints. Methode et resultats. *Bulletin et Memoires de la Societe d'Anthropologie de Paris* n.s., I: 95–118.

Henrich, J., S. J. Heine, and A. Norenzayan. 2010. The weirdest people in the world? *Behavioral and Brain Sciences* 33: 61–135.

Henshilwood, C. S., F. d'Errico, K. L. van Niekerk, Y. Coquinot, Z. Jacobs, S. E. Lauritzen, et al. 2011. A 100,000 year old ochre processing workshop at Blombos Cave, South Africa. *Science* 334: 219–221.

Herdt, G. H., ed. 1993. *Ritualized Homosexuality in Melanesia*. Berkeley: University of California Press.

Heyder, K., B. Suchan, and I. Daum. 2004. Cortico-subcortical contributions to executive control. *Acta Psychologia* 115: 271–289.

Higham, T., R. Jacobi, M. Julien, F. David, L. Basel, R. Wood, et al. 2010. Chronology of the Grotte du Renne (France) and implications for the context of ornaments and human remains within the Chatelperronian. *PNAS* 107: 20234–20239.

Hiiemae, K. M., J. B. Palmer, S. W. Medicis, J. Hegener, B. S. Jackson, and D. E. Lieberman, 2002. Hyoid and tongue movements in speaking and eating. *Archives of Oral Biology* 47: 11–27.

Hillenbrand, J. L., A. Getty, M. J. Clark, and K. Wheeler. 1995. Acoustic characteristics of American English vowels. *Journal of the Acoustical Society of America* 97: 3099–3111.

Hochstadt, J., H. Nakano, P. Lieberman, and J. Friedman. 2006. The roles of sequencing and verbal working memory in sentence comprehension deficits in Parkinson's disease. *Brain and Language* 97: 243–257.

Horning, J. J. 1969. A study of grammatical inference. PhD dissertation, Stanford University.

Howells, W. W. 1989. Skull shapes and the map: craniometric analyses in the dispersion of modern Homo. In *Papers of the Peabody Museum of Archaeology and Ethnology*, vol. 79. Cambridge, MA: Harvard University.

Humboldt, W. von. [1835] 1988. *On Language: The Diversity of Human Language-Structure and Its Influence on the Mental Development of Mankind*, tr. P. Heath. Cambridge, UK: Cambridge University Press.

Iwatsubo, T., S. Kuzuhara, A. Kanemitsu, H. Shimada, and Y. Toyokura. 1990. Corticofugal projections to the motor nuclei of the brainstem and spinal cord in humans. *Neurology* 140: 309–312.

Jackendoff, R. 1994. *Patterns in the Mind: Language and Human Nature*. New York: Basic Books.

Jellinger, K. 1990. New developments in the pathology of Parkinson's disease. In *Advances in Neurology*, vol. 53, *Parkinson's Disease: Anatomy, Pathol-*

ogy, and Therapy, ed. M. B. Streifler, A. D. Korezyn, J. Melamed, and M.B.H. Youdim. New York: Raven, 1–15.

Jeong, J. H., C. Jay, J. C. Kwon, J.Chin, S. J.Yoon, and D. L. Na, 2002. Globus pallidus lesions associated with high mountain climbing. *Journal of Korean Medical Sciences* 17: 861–863.

Jerison, H. J. 1973. *Evolution of the Brain and Intelligence*. New York: Academic Press.

Jin, X., and R. M. Costa. 2010. Start/stop signals emerge in nigrostriatal circuits during sequence learning. *Nature* 466: 457–462.

Johanson, D., and E. Maitland. 1981. *Lucy: The Beginnings of Humankind*. New York: Simon and Schuster.

Jones, D. 2010. A WEIRD view of human nature skews psychologists studies. *Science*: 328, 1627.

Joshua, M., A. Adler, R. Mitelman, E. Vaadia, and H. Bergman. 2008. Midbrain dopaminergic neurons and striatal cholinergic interneurons encode the difference between reward and aversive events at different epochs of probabilistic classical conditioning trials. *Journal of Neuroscience* 28: 11673–11684.

Jurgens, U. 2002. Neural pathways underlying vocal control. *Neuroscience and Biobehavioral Reviews* 26: 235–258.

Kagan, J. 1981. *The Second Year: The Emergence of Self-Awareness*. Cambridge, MA: Harvard University Press.

Kagan, J., J. S. Reznick, and N. Snidman. 1988. Biological bases of childhood shyness. *Science* 240: 167–171.

Kaminski, J., J. Call, and J. Fischer. 2004. Word learning in a domestic dog: evidence for "fast mapping." *Science* 304: 1682–1683.

Kegl, J., A. Senghas, and M. Coppola. 1999. Creation through contact: sign language emergence and sign language change in Nicaragua. In *Language Creation and Language Change: Creolization, Diachrony, and Development*, ed. M. Degraff. Cambridge, MA: MIT Press, 179–237.

Kelley, R. C. 1980. *Etoro Social Structure: A Study in Structural Contradiction*. Ann Arbor: University of Michigan Press.

Klein, R. G. 1999. *The Human Career*. 2nd ed. Chicago: Chicago University Press.

Knowlton, B. J., J. A. Mangels, and L. R. Squire. 1996. A neostratal learning system in humans. *Science* 273: 1399–1402.

Konopka, G., J. Bomarl, K. Winden, G. Coppola, Z. O. Jonsson, F. Gao, et al. 2009. CNS development genes by FOXP2. *Nature* 462: 213–217.

Kosslyn, S. M., A. Pascual-Leone, O. Felician, S. Camposano, J. P. Keenan, W. L. Thompson, et al. 1999. The role of area 17 in visual imagery: convergent evidence from PET and rTMS. *Science* 284: 167–170.

Kotz, S. A., S. Frisch, D. Y. von Cramon, and A. D. Friederici. 2003. Syntactic language processing: ERP lesion data on the role of the basal ganglia. *Journal of the International Neuropsychological Society* 9: 1053–1060.

Kourtzi, Z., H. H. Bulkthoff, M. Erb, and W. Grodd. 2002. Object-selective responses in the human motion area mt/mst. *Nature Neuroscience* 5: 17–18.

Krakauer, J. 1997. *Into Thin Air*. New York: Villard.

Krause, J., C. Lalueza-Fox, L. Orlando, W. Enard, R. E. Green, H. A. Burbano, et al. 2007. The derived FOXP2 variant of modern humans was shared with Neanderthals. *Current Biology* 17: 1908–1912.

Kuhl, P. K., K. A. Williams, F. Lacerda, K. N. Stevens, and B. Lindblom. 1992. Linguistic experience alters phonetic perception in infants by 6 months of age. *Science* 255: 606–608.

Kuoppamaki, M., K. Bhatia, and N. Quinn. 2002. Progressive delayed-onset dystonia after cerebral anoxic insult in adults. *Movement Disorders* 17: 1345–1349.

Kuypers, H. 1958. Corticobulbar connections to the pons and lower brainstem in man. *Brain* 81: 364–388.

Labov, W. M., Y.Yaeger, and R. Steiner. 1972. *A Quantitative Study of Sound Change in Progress*. Philadelphia: U.S. Regional Survey.

Ladefoged, P., and D. E. Broadbent. 1957. Information conveyed by vowels. *Journal of the Acoustical Society of America* 29: 98–104.

Lai, C. S., D. Gerrelli, A. P. Monaco, S. E. Fisher, and A. J. Copp. 2003. FOXP2 expression during brain development coincides with adult sites of pathology in a severe speech and language disorder. *Brain* 126: 2455–2462.

Lai, S. J., S. E. Fisher, J. A. Hurst, F. Vargha-Khadem, and A. P. Monaco. 2001. A forkhead-domain gene is mutated in a severe speech and language disorder. *Nature* 413: 519–523.

Lange, K.W., T. W. Robbins, C. D. Marsden, M. James, A. Owen, and G. M. Paul. 1992. L-dopa withdrawal in Parkinson's disease selectively impairs cognitive performance in tests sensitive to frontal lobe dysfunction. *Psychopharmacology* 107: 394–404.

Lartet, E. 1868. De quelques cas de progression organique vérifiables dans la succession des temps, géologiques sur des mammifères de même famille et de même genre. *Comptes Rendus de l'Académie des Sciences Paris* 66: 1119–1122.

Lashley, K. S. 1951. The problem of serial order in behavior. In *Cerebral Mechanisms in Behavior*, ed. L. A. Jeffress. New York: Wiley, 112–146.

Lehericy, S., M. Ducros, P.-F. Van De Moortele, C. Francois, L. Thivard, C. Poupon, et al. 2004. Diffusion tensor fiber tracking shows distinct corticostriatal circuits in humans. *Annals Neurology* 55: 522–527.

Leibnitz, G. W. [1765] 1916. *New Essays Concerning Human Understanding*, tr. A. M. Langley. Chicago: The Open Court Publishing Company.

Levelt, W. J. 1989. *Speaking: From Intention to Articulation*. Cambridge, MA: MIT Press.

Liberman, A. M., F. S. Cooper, D. P. Shankweiler, and M. Studdert-Kennedy. 1967. Perception of the speech code. *Psychological Review* 74: 431–461.

Lieberman, B. E. 2006. *Terrible Fate: Ethnic Cleansing in the Making of Modern Europe*. Chicago: Ivan R. Dee.

Lieberman, D. E. 2011. *The Evolution of the Human Head*. Cambridge, MA: Harvard University Press.

Lieberman, D. E., and J. J. Shea. 1994. Behavioral differences between archaic and modern humans in the Levantine Mousterian. *American Anthropologist* 96: 300–332.

Lieberman, D. E., and R. C. McCarthy. 1999. The ontogeny of cranial base angulation in humans and chimpanzees and its implications for reconstructing pharyngeal dimensions. *Journal of Human Evolution* 36: 487–517.

Lieberman, D. E., R. C. McCarthy, K. M. Hiiemae, and J. B. Palmer. 2001. Ontogeny of postnatal hyoid and laryngeal descent: implications for deglutition and vocalization. *Archives of Oral Biology* 46, 117–128.

Lieberman, M. R., and P. Lieberman. 1973. Olson's "projective verse" and the use of breath control as a structural element. *Language and Style* 5: 287–298.

Lieberman, P. 1963. Some measures of the fundamental periodicity of normal and pathologic larynges. *Journal of the Acoustical Society of America* 35: 344–353.

———. 1968. Primate vocalizations and human linguistic ability. *Journal of the Acoustical Society of America* 44: 1157–1164.

———. 1980. On the development of vowel production in young children. In *Child Phonology, Perception and Production*, ed. G. Yeni-Komshian and J. Kavanagh. New York: Academic Press, 113–142.

———. 1984. *The Biology and Evolution of Language*. Cambridge, MA: Harvard University Press.

———— . 2000. *Human Language and Our Reptilian Brain: The Subcortical Bases of Speech, Syntax, and Thought*. Cambridge MA: Harvard University Press.

———— . 2002. On the nature and evolution of the neural bases of human language. *Yearbook of Physical Anthropology* 45: 36–62.

———— . 2006 *Toward an Evolutionary Biology of Language*. Cambridge, MA: Harvard University Press.

———— . 2009. FOXP2 and human cognition. *Cell* 137: 800–802.

———— . 2012. Vocal tract anatomy and the neural bases of talking. *Journal of Phonetics* 40: 608–622.

Lieberman, P., and S. E. Blumstein. 1988. *Speech Physiology, Speech Perception, and Acoustic Phonetics*. Cambridge, UK: Cambridge University Press.

Lieberman, P., and E. S. Crelin, 1971. On the speech of Neanderthal man. *Linguistic Inquiry* 2: 203–222.

Lieberman, P., and R. M. McCarthy. 2007. Tracking the evolution of language and speech. *Expedition* 49: 15–20.

Lieberman, P., and S. B. Michaels. 1962. Some aspects of fundamental frequency and envelope amplitude as related to the emotional context of speech. *Journal of the Acoustical Society of America* 34: 922–927.

Lieberman, P., D. H. Klatt, and W. H. Wilson. 1969. Vocal tract limitations on the vowel repertoires of rhesus monkey and other nonhuman primates. *Science* 164: 1185–1187.

Lieberman, P., E. S. Crelin, and D. H. Klatt. 1972. Phonetic ability and related anatomy of the newborn, adult human, Neanderthal man, and the chimpanzee. *American Anthropologist* 74: 287–307.

Lieberman, P., J. Friedman, and L. S. Feldman. 1990. Syntactic deficits in Parkinson's disease. *Journal of Nervous and Mental Disease* 178: 360–365.

Lieberman, P., E. T. Kako, J. Friedman, G. Tajchman, L. S. Feldman, and E. B. Jiminez. 1992. Speech production, syntax comprehension, and cognitive deficits in Parkinson's disease. *Brain and Language* 43: 169–189.

Lieberman, P., B. G. Kanki, A. Protopapas, E. Reed, and J. W. Youngs. 1994. Cognitive defects at altitude. *Nature* 372: 325.

Lieberman, P., and S. B. Michaels. 1962. Some aspects of fundamental frequency and envelope amplitude as related to the emotional context of speech. *Journal of the Acoustical Society of America* 34: 922–927.

Lieberman, P., A. Morey, J. Hochstadt, M. Larson, and S. Mather. 2005. Mount Everest: a space-analog for speech monitoring of cognitive deficits and stress. *Aviation, Space and Environmental Medicine* 76: 198–207.

Liegeois, F., T. Baldeweg, A. Connelly, D. G. Gadian, M. Mishkin, and F. Vargha–Khadem. 2003. Language fMRI abnormalities associated with FOXP2 gene mutation. *Nature Neuroscience* 6: 1230–1237.

Lisker, L., and A. S. Abramson. 1964. A cross language study of voicing in initial stops: acoustical measurements. *Word* 20: 384–442.

Logothetis, N. K., J. Pauls, M. Augath, T. Trinath, and A. Oeltermann. 2001. Neurophysiological investigation of the basis of the fMRI signal. *Nature* 412: 150–157.

Long, C. A. 1969. The origin and evolution of mammary glands. *Biological Sciences* 19: 519–523.

Lubker, J., and T. Gay. 1982. Anticipatory labial coarticulation: experimental, biological, and linguistic variables. *Journal of the Acoustical Society of America* 71: 437–438.

MacLean, P. D. 1986. Neurobehavioral significance of the mammal-like reptiles (therapsids). In *The Ecology and Biology of Mammal-Like Reptiles*, ed. N. Hotton III, J. J. Roth, and E. C. Roth. Washington, DC: Smithsonian Institution Press, 1–21.

MacLean, P. D., and J. D. Newman. 1988. Role of midline frontolimbic cortex in the production of the isolation call of squirrel monkeys. *Brain Research* 450: 111–123.

Maess, B., S. Koelsch, T. C. Gunter, and A. D. Friederici. 2001. Music syntax is processed in the area of Broca: an MEG study. *Nature Neuroscience* 4: 540–545.

Maguire, E. A., K. Woollett, and H. J. Spiers. 2006. London taxi drivers and bus drivers: a structural MRI and neuropsychological analysis. *Hippocampus* 16: 1091–1101.

Mahajan, P. V., and B. A. Bharucha. 1994. Evaluation of short neck: percentiles and linear correlations with height and sitting height. *Indian Pediatrics* 31: 1193–1203.

Marie, P. 1926. *Traveaux et mémoires*. Paris: Masson.

Marin, O., W. J. Smeets, and A. Gonzalez. 1998. Evolution of the basal ganglia in tetrapods: a new perspective based on recent studies in amphibians. *Trends in Neurosciences* 21: 487–494.

Marsden, C. D., and J. A. Obeso. 1994. The functions of the basal ganglia and the paradox of sterotaxic surgery in Parkinson's disease. *Brain* 117: 877–897.

Martin, A., and L. L. Chao. 2001. Semantic memory and the brain: structure and processes. *Current Opinion in Neurobiology* 11: 194–201.

Martin, A., J. V. Haxby, F. M. Lalonde, C. L. Wiggs, and L. G. Ungerleider. 1995. Discrete cortical regions associated with knowledge of color and knowledge of action. *Science* 270: 102–105.

Mateer, C. J., and G. A. Ojemann. 1983. Thalamic lesions in language and memory. In *Language Functions and Brain Organization*, ed. S. Segalowitz. New York: Academic Press, 171–191.

Mayr, E. 1982. *The Growth of Biological Thought*. Cambridge, MA: Harvard University Press.

McBrearty, S., and A. S. Brooks. 2000. The revolution that wasn't: a new interpretation of the origin of modern human behavior. *Journal of Human Evolution* 39: 453–563.

McCarthy, R. C., and D. S. Strait. 2005. Morphological constraints on hominin speech production. *Paleoanthropology* 3: AO2.

McElligott, A. G., M. Birrer, and E. Vannoni. 2006. Retraction of the mobile descended larynx during groaning enables fallow bucks (*Dama dama*) to lower their formant frequencies. *Journal of Zoology* 270, 340–345.

McNeill, D. 1985. So you think gestures are nonverbal? *Psychological Review* 92: 350–371.

Melchner, L. von, S. L. Pallas, and M. Sur. 2000. Visual behavior mediated by retinal projections diverted to the auditory pathway. *Nature* 404: 871–876.

Meyer, M., M. Kircher, M-T Gansauge, H. Li, F. Racimo, S. Mallick, J. G. Schraiber, F. Jay, et al. 2012. A high-coverage genome sequence from an archaic Denisovan individual. *Science* 337: published online 30, August 2012.

Miller, E. K, and J. D. Wallis. 2009. Executive function and higher order cognition: definition and neural substrates. In *Encyclopedia of Neuroscience*, ed. L. Squire, vol. 4. New York: Springer, 99–104.

Mirenowicz, J., and W. Schultz. 1996. Preferential activation of midbrain dopamine neurons by appetitive rather than aversive stimuli. *Nature* 379: 449–451.

Monchi, O., M. Petrides, K. Petre, K. J. Worsley. and A. Dagher. 2001. Wisconsin card sorting revisited: distinct neural circuits participating in different stages of the test as evidenced by event-related functional magnetic resonance imaging. *The Journal of Neuroscience* 21: 7739–7741.

Monchi, O., M. Petrides, A. P. Strafella, K. J. Worsely, and A. Doyon. 2006a. Functional role of the basal ganglia in the planning and execution of actions. *Annals of Neurology* 59: 257–264.

Monchi, O., J. H. Ko, and A. P. Strafella. 2006b. Striatal dopamine release during performance of executive function: a [11C] raclopride PET study. *Neuroimage* 33: 907–912.

Monchi, O., M. Petrides, B. Meja-Constain, and A. P. Strafella. 2007. Cortical activity in Parkinson disease during executive processing depends on striatal involvement. *Brain* 130: 233–244.

Moore, L. G., S. Niermeyer, and S. Zamudio. 1998. Human adaptation to high altitude: regional and life-cycle perspectives. *Yearbook of Physical Anthropology* 41: 25–61.

Mouse Genome Sequencing Consortium. 2002. Initial sequencing and comparative analysis of the mouse genome. *Nature* 420: 520–562.

Muller, J. 1848. *The Physiology of the Senses, Voice and Muscular Motion with the Mental Faculties*, tr. W. Baly. London: Walton and Maberly.

Naeser, M. A., M. P. Alexander, N. Helms-Estabrooks, H. L. Levine, S. A. Laughlin, and N. Geschwind. 1982. Aphasia with predominantly subcortical lesion sites: description of three capsular/putaminal aphasia syndromes. *Archives of Neurology* 39: 2–14.

National Safety Council. 2010. *Highlights from Injury Facts*. Available at www.nsc.org/news_resources/injury_and_death_statistics/Pages/HighlightsFromInjuryFacts.aspx; accessed May 2012.

Natsopoulos, D., G. Grouios, S. Bostantzopoulou, G. Mentenopoulos, Z. Katsarou, and J. Logothetis. 1993. Algorithmic and heuristic strategies in comprehension of complement clauses by patients with Parkinson's disease. *Neuropsychologia* 31: 951–964.

Nauta, W.J.H., and P. A. Gygax. 1954. Silver impregnation of degenerating axons in the central nervous system: a modified technic. *Stain Technology* 29: 91–93.

Nearey, T. 1978. *Phonetic Features for Vowels*. Bloomington: Indiana University Linguistics Club.

Negus, V. E. 1949. *The Comparative Anatomy and Physiology of the Larynx*. New York: Hafner.

Newman, A. J., D. Bavelier, D. Corina, P. Jezzard, and H. J. Neville. 2002. A critical period for right hemisphere recruitment in American Sign Language processing. *Nature Neuroscience* 5: 76–80.

Newman, J. D. 1985. The infant cry of primates: an evolutionary perspective. In *Infant Crying: Theoretical and Research Perspectives*, ed. B. M. Lester and C. F. Zachariah Boukydis. New York: Plenum, 307–323.

O'Brian, P. 1972. *Post Captain*. New York: W. W. Norton.

O'Toole, A. J., F. Jiang, H. Abdi, and J. V. Haxby. 2005. Partially distributed representations of objects and faces in ventral temporal cortex. *Journal of Cognitive Neuroscience* 17: 580–590.

Pal, D. K, W. Li, T. Clarke, P. Lieberman, and L. J. Strug. 2010. Pleiotropic effects of the 11p13 locus on developmental verbal dyspraxia and EEG centrotemporal sharp waves. *Genes, Brain and Behavior* 9: 1004–1012.

Palmer, J. B., N. J. Rudin, G. Lara, and A. W. Crompton. 1992. Coordination of mastication and swallowing. *Dysphagia* 7: 187–200.

Patel, A. D. 2008. *Music, Language and the Brain.* New York: Oxford University Press.

Pembrey, M. E., L. O. Bygren, G. Kaati, S. Edvinsson, K. Northstone, M. Sjöström, J. Golding, and ALSPAC Study Team. 2006. Sex-specific, male-line transgenerational responses in humans. *European Journal of Human Genetics* 14: 159–166.

Peterson, G. E., and H. L. Barney. 1952. Control methods used in a study of the vowels. *Journal of the Acoustical Society of America* 24: 175–184.

Petrides, M. 2005. Lateral prefrontal cortex: architectonic and functional organization. *Philosophical Transactions of the Royal Society B* 360: 781–795.

Pickett, E. R., E. Kuniholm, A. Protopapas, J. Friedman, and P. Lieberman. 1998. Selective speech motor, syntax and cognitive deficits associated with bilateral damage to the putamen and the head of the caudate nucleus: a case study. *Neuropsychologia* 36: 173–188.

Pinhasi, R., T.F.G. Higham, L. V. Golovanova, and V. B. Doronichev. 2011. Revised age of late Neanderthal occupation and the end of the Middle Paleolithic in the northern Caucasus. *Proceedings of the National Academy of Sciences USA* 108: 8611–8618.

Pinker, S. 1994. *The Language Instinct: How The Mind Creates Language.* New York: William Morrow.

———. 1998. *How the Mind Works.* New York: Norton.

———. 2002. *The Blank Slate.* New York: Norton.

———. 2007. *The Stuff of Thought.* New York: Viking.

———. 2011. *The Better Angels of Our Nature: Why Violence Has Declined.* New York: Viking.

Podos, J. 2001. Correlated evolution of morphology and vocal signal structure in Darwin's finches. *Nature* 409: 185–188.

Polich, L. 2006. *The Emergence of the Deaf Community in Nicaragua: With Sign Language You Can Learn So Much.* Washington, DC: Gallaudet University Press.

Postle, B. R. 2006. Working memory as an emergent property of the mind and brain. *Neuroscience* 139: 23–38.

Prescott, W. H. [1843] 1925. *History of the Conquest of Mexico.* New York: Random House.

Preuss, T. M., M. Caceres, M. C. Oldham, and D. H. Geschwind. 2004. Human brain evolution: insights from microarrays. *Nature Reviews: Genetics* 5: 850–860.

Provost, J. S., M. Petrides, and O. Monchi. 2010. Caudate nucleus involvement in the monitoring of events within working memory. *European Journal of Neuroscience* 32: 873–880.

Provost, J.-S., M. Petrides, F. Simard, and O Monchi. 2011. Investigating the long-lasting residual effect of a set-shift on frontostriatal activity. *Cerebral Cortex* doi: 10.1093/cercor/bhr358.

Ptak, S., W. Enard, V. Wiebe, I. Hellmann, J. Krause, M. Lachmann, and S. Paabo. 2009. Linkage disequilibrium extends across putative selected sites in FOXP2. *Molecular Biology and Evolution* 26: 2181–2184.

Pullum, G. 1991. *The Great Eskimo Vocabulary Hoax and Other Irreverent Essays on the Study of Language.* Chicago: University of Chicago Press.

———. 2011. Ling-Lang website posting: remarks by Noam Chomsky in London. *Linguist List* 22.4631, November 19, 2011.

Reich, D., R. E. Green, M. Kircher, J. Krause, N. Patterson, E. Y. Durand, et al. 2010. Genetic history of an archaic hominen group from Denisova cave in Siberia. *Nature* 468: 1053–1060.

Reimers-Kipping, S., S. Hevers, S. Paabo, and W. Enard. 2011. Humanized FOXP2 specifically affects cortico-basal ganglia circuits. *Neuroscience* 175: 75–84.

Riel-Salvatore, J. 2010. A niche construction perspective on the Middle-Upper Paleolithic transition in Italy. *Journal of Archeological Method and Theory* 17: 323–355.

Rissman, J., J. C. Eliassen, and S. E. Blumstein. 2003. An event-related fMRI investigation of implicit semantic priming. *Journal of Cognitive Neuroscience* 15: 1160–1175.

Robertson, R. L., L. Ben-Sira, P. D. Barnes, R. V. Mulkern, C. D. Robson, S. E. Maier, et al. 1999. MR line-scan diffusion-weighted imaging of term neonates with perinatal brain ischemia. *American Journal of Neuroradiology* 20: 1658–1670.

Sachs, J., P. Lieberman, and D. Erickson, 1972. Anatomical and cultural determinants of male and female speech. In *Language Attitudes: Current Trends and Prospects*, Monograph No. 25, Georgetown University Monograph Series in Language and Linguistics. Washington, DC: Georgetown University.

Sanes, J. N., J. P. Donoghue, V. Thangaraj, R. R. Edelman, and S. Warach. 1999. Shared neural substrates controlling hand movements in human motor cortex. *Science* 268: 1775–1777.

Savage-Rumbaugh, S., D. Rumbaugh, and K. McDonald. 1985. Language learning in two species of apes. *Neuroscience and Biobehavioral Reviews* 9: 653–665.

Scarf, D., H. Hayne, and P. Colombo. 2011. Pigeons on par with primates in numerical competence. *Science* 334: 1664.

Scott, R. B., J. Harrison, C. Boulton, J. Wilson, R. Gregory, S. Parkin, et al. 2002. Global attentional-executive sequelae following surgical lesions to globus pallidus interna. *Brain* 125: 562–574.

Semendeferi, K., H. Damasio, R. Frank, and G. W. van Hoesen. 1997. The evolution of the frontal lobes: a volumetric analysis based on three-dimensional reconstructions of magnetic resonance scans of human and ape brains. *Journal of Human Evolution* 32: 375–378.

Semendeferi, K., A. Lu, N. Schenker, and H. Damasio. 2002. Humans and apes share a large frontal cortex. *Nature Neuroscience* 5: 272–276.

Schulz, G. M., M. Varga, K. Jeffries, C. L. Ludlow, and A. R. Braun. 2005. Functional neuroanatomy of human vocalization: an H 250 PET study. *Cerebral Cortex* 15: 1835–1847.

Shenhav, A., and J. D. Greene. 2010. Moral judgments recruit domain-general valuation mechanisms to integrate representations of probability and magnitude. *Neuron* 67: 667–677.

Shu, W., H. Yang, L. Zhang, M. M. Ju, and E. E. Morrisey. 2001. Characterization of a new subfamily of winged-helix/forkhead (Fox) genes that are expressed in the lungs and act as transcriptional repressors. *Journal of Biological Chemistry* 276: 27488–27497.

Shultz, S., C. Opie, and Q. D. Atkinson. 2011. Stepwise evolution of stable sociality in primates. *Nature* 479: 219–222.

Simard, F., Y. Joanette, M. Petrides, T. Jubault, C. Madjar, and O. Monchi. 2011. Fronto-striatal contributions to lexical set-shifting. *Cerebral Cortex* 21: 1084–1093.

Simonson, T. S., Y. Yang, C. D. Huff, H. Yun, G. Qin, D. J. Witherspoon, et al. 2010. Genetic evidence for high-altitude adaptation in Tibet. *Science* 329: 71–74.

Slocombe, K., T. Kaller, J. Call, and K. Zuberbühler. 2010. Chimpanzees extract social information from agonistic screams. *PLoS ONE* 5(7): e11473.

Smith, A., and L. Goffman. 1998. Stability of speech movement sequences in children and adults. *Journal of Speech, Language and Hearing Research* 41: 18–30.

Smith, B. L. 1978. Temporal aspects of English speech production: a developmental perspective. *Journal of Phonetics* 6: 37–68.

Spoor, F., M. G. Leakey, P. N. Gathogo, F. H. Brown, S. C. Antón, I. McDougall, et al. 2007. Implications of new early Homo fossils from Ileret, east of Lake Turkana, Kenya. *Nature* 448: 688–691.

Spurzheim, J. K. 1815. *The Physiognomical System of Drs. Gall and Spurzheim.* London: Printed for Baldwin, Cradock, and Joy.

Stephan, H., H. Frahm, and G. Baron. 1981. New and revised data on volumes of brain structures in insectivores and primates. *Folia Primatologia* 35: 1–29.

Stevens, K. N. 1972. Quantal nature of speech. In *Human Communication: A Unified View*, ed. E. E. David Jr. and P. B. Denes. New York: McGraw Hill, 51–66.

Stuss, D. T., and D. F. Benson. 1986. *The Frontal Lobes.* New York: Raven.

Swift, J. [1726] 1970. *Gulliver's Travels.* New York: Norton.

Swaminath, P. V., M. Ragothaman, U. B. Muthane, S.A.H. Udupa, S. L Rao, and S. S. Giovindappa. 2006. Parkinsonism and personality changes following an acute hypoxic insult during mountaineering. *Movement Disorders* 21: 1296–1297.

Takemoto, H. 2001. Morphological analyses of the human tongue musculature for three-dimensional modeling. *Journal of Speech, Language and Hearing Research* 44: 95–107.

———. 2008. Morphological analyses and 3D modeling of the tongue musculature of the chimpanzee (*Pan troglodytes*). *American Journal of Primatology* 70: 966–975.

Talairach, J., and Tournoux, P. 1988. *Co-planar Stereotaxic Atlas of the Human Brain.* Stuttgart: Thieme.

Taunton, M. 1982. Aesthetic responses of young children to the visual arts: a review of the literature. *Journal of Aesthetic Education* 16: 93–109.

Templeton, A. R. 2002. Out of Africa again and again. *Nature* 416: 45–51.

Teramatsu, I., and S. A. White. 2006. FoxP2 regulation during undirected singing in adult songbirds. *The Journal of Neuroscience* 26: 7390–7394.

Terao, S.-I., M. Li, Y. Hashizume, Y. Osano, T. Mitsuma, and G. Sobue. 1997. Upper motor neuron lesions in stroke patients do not induce anterograde transneuronal degeneration in spinal anterior horn cells. *Stroke* 28: 2553–2556.

Tishkoff, S. A., F. A. Reed, A. Ranciarol, B. F. Voight, C. C. Babbitt, J. S. Silverman, et al. 2007. Convergent adaptation of human lactose persistence in Africa and Europe. *Nature Genetics* 39: 31–39.

Tomasello, M. 2003. *Constructing a Language: A Usage-based Theory of Language Acquisition.* Cambridge, MA: Harvard University Press.

———. 2009. *Why We Cooperate.* Cambridge, MA: MIT Press.

Toth, N., and K. Schick. 1993. Early stone industries. In *Tools, Language and Cognition in Human Evolution,* ed. K. R. Gibson and T. Ingold. Cambridge, UK: Cambridge University Press, 346–362.

Truby, H. L., J. F Bosma, and J. Lind. 1965. *Newborn Infant Cry.* Upsalla: Almquist and Wiksell.

Tyson, E. 1699. *Orang-outang, sive, Homo sylvestris; or, The anatomy of a pygmie compared with that of a monkey, an ape, and a man.* London: Printed for Thomas Bennet and Daniel Brown.:480–485.

Vargha-Khadem, F., K. E. Watkins, K. Alcock, P. Fletcher, and R. Passingham. 1995. Praxic and nonverbal cognitive deficits in a large family with a genetically transmitted speech and language disorder. *Proceedings of the National Academy of Sciences USA* 92: 930–933.

Vargha-Khadem, F., K. E. Watkins, C. Price, J. J. Ashburner, K. J. Alcock, A. Connelly, et al. 1998. Neural basis of an inherited speech and language disorder. *Proceedings of the National Academy of Sciences USA* 95: 12695–12700.

Vargha-Khadem, F., D. G. Gadian, A. Copp, and M. Mishkin. 2005. FOXP2 and the neuroanatomy of speech and language. *Nature Reviews, Neuroscience* 6: 131–138.

Wade, N. 2009. *The Faith Instinct.* New York: Penguin.

Wang, J., H. Roa, G. S. Wetmore, P. M. Furlan, M. Korczykowski, D. F. Dinges, and J. A. Detre. 2005. Perfusion functional MRI reveals cerebral blood flow pattern under psychological stress. *Proceedings of the National Academy of Sciences USA* 102: 17804–17809.

Watkins, K. E., F. Vargha-Khadem, J. Ashburner, R. E. Passingham, A. Connelly, K. J. Friston, et al. 2002. MRI analysis of an inherited speech and language disorder: structural brain abnormalities. *Brain* 125: 465–478.

Wechsler, D. 1976. *Wechsler Intelligence Scale for Children, Revised*. Windsor, UK: Psychological Corporation.

———. 1986. *Wechsler Adult Intelligence Scale, Revised*. Windsor, UK: Psychological Corporation.

Wernicke, C. [1874] 1967. The aphasic symptom complex: a psychological study on a neurological basis. In *Proceedings of the Boston Colloquium for the Philosophy of Science*, vol. 4, ed. R. S. Cohen and M. W. Wartofsky. Dordrecht: Reidel.

Westermann, E. B. 2005. *Hitler's Police Battalions: Enforcing Racial War in the East*. Lawrence: University Press of Kansas.

Whorf, B. L. 1956. *Language, Thought, and Reality: Selected Writings of Benjamin Lee Whorf*. Cambridge, MA: Technology Press of Massachusetts Institute of Technology.

Wrangham, R., and D. Peterson, 1996. *Demonic Males: Apes and the Origins of Human Violence*. New York: Houghton Mifflin.

Wynne-Edwards, V. C. 1962. *Animal Dispersion in Relation to Social Behavior*. New York, Hafner.

Yi, X., Y. Liang, E. Huerta-Sanchez, X. Jin, Z. K. Cuo, J. E. Pool, et al. 2010. Sequencing of 50 human exomes reveals adaptation to high altitude. *Science* 329: 75–78.

Zhang, J. 2003. Evolution of the human ASPM gene, a major determinant of brain size. *Genetics* 165: 2063–2070.

Index

Abarbanell, L., 184–85
Abi-Rached, L. M., 152, 154
Abramson, A., 41
Acheulian tools, 126–27
acoustics: ape speech and, 134–36;
    Bell Telephone Laboratories and,
    134, 136; fundamental frequency
    (F) and, 39, 81, 192, 200; gender
    recognition and, 199–202; Haskins
    Laboratories and, 134–35; music
    and, 14, 38, 77, 81, 93, 148, 154,
    173, 201; pitch and, 39, 81, 135,
    192, 200–201; quantal vowels and,
    137–38 (see also speech); Sound
    Spectrograph and, 41, 134–35;
    supralaryngeal vocal tract (SVT)
    and, 200
adaptation: adult lactose tolerance and,
    68–69, 166; archaeology and, 88;
    brain design and, 2–3, 25; changes
    in ecosystems and, 73; evolution
    and, 74–75, 78; exaptation and, 76;
    extreme altitudes and, 70–71, 166;
    language and, 72–73, 90–91; pread-
    aptation and, 76

*Adapted Mind, The: Evolutionary Psy-
    chology and the Generation of Cul-
    ture* (Cosmides and Toobey), 158
Adeylott, J., 84
Africa, 67, 69–70, 87, 114, 122, 147,
    150–53, 156–57
Aldridge, W., 173–74
Alexander, G. E., 34, 111
alpha males, 195–96
Altmann, J., 161
Altmann, S., 161
altruism, 3, 185–86, 194
Alzheimer's dementia, 42, 89
American Association for the Advance-
    ment of Science (AAAS), 100, 103
American Sign Language (ASL), 91,
    123, 176–77
Amnesty International (AI), 180
Amunts, K. A., 16
amygdala, 9–10
anatomy: ape vocalization and, 133–36;
    brain size and, 2, 11, 83–88, 113,
    118; neuroimaging studies and, 11–
    18; speech and, 131f, 134–43; swal-
    lowing and, 64, 79, 129–32, 144–46;

231

Blair, R.J.R., 10

blood oxygen level depletion (BOLD) signal, 13

Boehm, C., 195

Boesch, C., 99, 124

Book of the Five Rings, The (Musashi), 56

Boston Language Development Conference, 100

Boston University, 174

Botticelli, S., 144

Bouhuys, A., 70

Bowerman, M., 169

Brainard, M. S., 112, 202–3

brain damage: anterior cingulate cortex (ACC) and, 35–36; aphasia and, 98–99; basal ganglia and, 44, 98; brain design and, 27–36, 47; Broca studies and, 27; CM (patient) and, 41–42, 44, 47, 80; comas and, 34, 41, 98; language and, 98; lesions and, 99 (*see also* lesions); mechanism effects and, 16, 18; motor control and, 76; Parkinson disease and, 99; speech and, 98; strokes and, 27–29, 34–35, 97; Tan (patient) and, 27–28

brain design: adaptation and, 2–3, 25; anterior cingulate cortex (ACC) and, 35–36; axons and, 18–20, 120; basal ganglia and, 28–37, 41, 44–58; behavior and, 26–27, 34–36, 48; brain damage and, 27–36, 47; caudate nucleus and, 33f, 34, 41, 46, 52–54, 57–58; cognition and, 31–37, 41–59; cortical-basal ganglia-cortical circuits and, 32–36; dendrites and, 18–20, 36, 111, 120; emotion and, 32–33, 35f, 36–37, 54; executive control and, 44, 49–54; fissures and, 10, 11f; functional magnetic resonance imaging (fMRI) and, 27, 36, 51–55, 58–59; hemispheres of, 10, 16, 97; hippocampus and, 54; information and, 27, 33f, 34–36, 50–53; learning and, 37–38, 41, 51–52, 55–59; local

operations and, 30, 32, 36–37, 44, 51–57; mathematics and, 25–27, 59; memory and, 42, 50–54, 57; modular theory and, 54, 76, 164, 208; monkeys and, 50, 58; motor control and, 32, 35f, 36–37, 40, 44–45; neural circuits and, 28–36, 45, 49, 51, 53–55; positron emission tomography (PET) scans and, 27, 36, 51, 53–54; prefrontal cortex and, 35–36, 41–55, 58; putamen and, 33f, 34, 41, 46, 53–54, 57; speech and, 27–32, 35, 37–47, 55, 58–59; survival and, 30–31, 48–49, 61, 63; thalamus and, 30, 33f, 34–36, 52, 55

brain mechanisms, 208; amygdala and, 9–10; anterior cingulate cortex (ACC) and, 9, 22; assembly lines and, 5–6; behavior and, 2–5, 13, 19–24; computational architecture of, 3; cortical malleability and, 18; emotion and, 9, 22; evolutionary psychology and, 2–3, 5, 24; functional architecture and, 4; hippocampus and, 23; information and, 8, 14–15, 18–19, 23; limited understanding of, ix–x; local operations and, x, 4, 21–22, 30, 32, 36–37, 44, 51–57, 77, 94, 100, 110, 190; mathematics and, 5–8; memory and, 8, 19–23; modular theory and, x, 3–7, 54, 76, 164, 208; monkeys and, 16; motor control and, 3, 5, 9, 21–23; neural circuits and, 3–4, 9, 14, 21–23, 190; neuroimaging and, 11–18; phrenology and, 6–8, 22; plasticity and, 18; positron emission tomography (PET) scans and, 12–14; prefrontal cortex and, 8–10, 11f, 16, 17f; religion and, 3, 8–9, 22; speech and, 5, 14, 18, 23–24; supercharged, 82–83, 112–18, 120; synapses and, 18–21

brain size, 2, 84; apes and, 83, 86; ASPM gene and, 113; chimpanzees and, 11, 85; Darwin and, 86–87; increasing of, 86–88, 113, 118; Neanderthals and, 152; struggle for existence and, 87

breast cancer, ix

Broadbent, D. E., 139

Broca, P., 27–28, 32

Broca's area: aphasia and, 29–30, 32, 40, 45, 189; basal ganglia and, 9, 15, 22–23; entrenched genes and, 191–93; fossils and, 26; language and, 26–32, 189; morals and, 30; neural circuits and, 107; striatum and, 107

Brodmann, K., 16–18, 94

Brodmann's areas (BAs), 16–18, 28

Bronowski, J., 90

Brooks, A. S., 150

Browne, J., 64, 116

Brown University, 31

Buckner, R., 89

Burling, R., 169

Button, J., 122

Cabela's, 157

Calvin, W., 87

cancer, ix

Candide (Voltaire), 75

cardiovascular disease, 203

Carew, T. J., 21, 111

Carre, R., 138, 145

cars, 117–18

Cartesian Linguistics (Chomsky), 170

cats, 129–30

caudate nucleus: brain design and, 33f, 34, 41, 46, 52–54, 57–58; FOXP2 gene and, 106–8, 112; KE family and, 106–8; lesions and, 46, 105; Monchi group and, 95

Cavness, B., 133

cerebellum, 88, 108

Chagall, M., 148

Chao, L. L., 88–89

Chapais, B., 159–61

Chatelperronian culture, 153–54

cheater-detectors, xi, 158–59

Chie, U., 46, 105

Chimpanzee Politics: Power and Sex among Apes (de Waal), 195–96

chimpanzees, 131f, 161; alpha-male status and, 195–96; American Sign Language (ASL) and, 123, 177; as baseline, 123–24; behavior and, 148; brain size and, 2, 11, 85; calls of, 134–36; Chimpanzee Sequencing and Analysis Consortium and, 109; cognition and, 195–96; cost-benefit analysis and, 195–96; creativity and, 147–48; Darwin and, x; fission-fusion society and, 194; food and, 194, 198; FOXP2$_{human}$ gene and, 152; Frido, 196; gender and, 201; genes shared with humans, 107; genetics and, ix, 24, 82–83, 85–86, 91, 95–96, 99–100, 107, 109–10, 113–15, 117–18, 190, 194–95; Gombe, 124, 194, 196, 198; groupings of, 194–97; human superiority over, 30, 95–96; hunting parties and, 198; Kanzi, 123; Leakeys and, 124–25; learning and, 127, 161; mirror-test and, 198; morals and, 194–98; neural circuits and, 76–80, 95–96; prefrontal cortex and, 85; reading and, 95–96; sex and, 160; simple tools and, 124–25, 147–48, 195; speech and, 95, 134–36; supercharged brains and, 82; syntax and, 123; tutoring and, 127; warfare and, 195–96; Washoe, 91, 123, 177; Wilkie, 196

Chimpanzees of Gombe, The (Goodall), 124

China, 197–98

Chinese, 71–73, 143, 164–65, 167, 204

choking, 23, 130–32, 146

Chomsky, C., 169

Chomsky, N.: behavior and, 163, 165, 167–68, 171–73; bird brains and, 168; creationist linguistics and, 162–66; Darwin and, 164–66; determinism and, x–xi; Everett paper and, 205–6; evolutionary psychology and,